An IEEE 802.11 Based
Wireless Mesh Disaster Recovery System
with Lifetime Enhancement

An IEEE 802.11 Based Wireless Mesh Disaster Recovery System with Lifetime Enhancement

System Design, Hardware Requirements
and Performance Evaluation

Ein IEEE 802.11 basiertes Wireless Mesh Disaster Recovery System mit Lebenszeitverlängerung

Systementwurf, Hardwareanforderungen
und Performancebewertung

Der Technischen Fakultät
der Friedrich-Alexander-Universität
Erlangen-Nürnberg
zur

Erlangung des Doktorgrades Dr.-Ing.

vorgelegt von

Christopher Hepner
aus Illertissen

Bibliografische Information der Deutschen Nationalbibliothek
Die Deutsche Nationalbibliothek verzeichnet diese Publikation in der
Deutschen Nationalbibliografie; detaillierte bibliografische Daten
sind im Internet über http://dnb.d-nb.de abrufbar.
1. Aufl. - Göttingen: Cuvillier, 2019
Zugl.: Erlangen-Nürnberg, Univ., Diss., 2019

Als Dissertation genehmigt von
der Technischen Fakultät
der Friedrich-Alexander-Universität Erlangen-Nürnberg
Tag der mündlichen Prüfung: 25.01.2019

Vorsitzender des Promotionsorgans: Prof. Dr.-Ing. Reinhard Lerch

Gutachter: Prof. Dr.-Ing. Dr.-Ing. habil. Robert Weigel
              Prof. Dr. rer. nat. Roland Münzner
              Prof. Dr.-Ing. Robert Fischer

© CUVILLIER VERLAG, Göttingen 2019
Nonnenstieg 8, 37075 Göttingen
Telefon: 0551-54724-0
Telefax: 0551-54724-21
www.cuvillier.de

ISBN 978-3-7369-7052-6
eISBN 978-3-7369-6052-7

Danksagung

An dieser Stelle möchte ich mich bei einigen Personen bedanken, die mich während der Entstehung dieser Arbeit unterstützt haben.

Mein erster Dank gilt Herrn Prof. Dr.-Ing. Dr.-Ing. habil. Robert Weigel für die Betreuung meiner Arbeit am Lehrstuhl für Technische Elektronik der Friedrich-Alexander-Universität Erlangen-Nürnberg. Vielen Dank für die konstruktiven Tipps und der Möglichkeit des selbstverantwortlichen Arbeitens.

Mein besonderer Dank gilt Herrn Prof. Dr. rer. nat. Roland Münzner für die hervorragende fachliche Betreuung und für die Möglichkeit, in Kooperation mit der Hochschule Ulm, unter seiner Leitung promovieren zu dürfen. Danke für die unzähligen konstruktiven Diskussionen in mehrstündigen Meetings und das entgegengebrachte Vertrauen.

Vielen Dank an Herrn Prof. Dr.-Ing. Robert Fischer (Universität Ulm) für die Übernahme des Drittgutachtens.

Ebenfalls ein besonderer Dank gilt den Mitarbeitern des Instituts für Kommunikationstechnik für die sehr gute Zusammenarbeit am Institut, insbesondere die aufmunternden Gespräche in der Mittagsrunde. Besonderer Dank an Stefan Fuchs für die Unterstützung bei allen Fragen zur Laborinfrastruktur. Ebenfalls Danken möchte ich Jannik Maier für die Unterstützung im Endspurt der Arbeit.

Vielen Dank auch für die kollegiale Zusammenarbeit den Mitarbeitern des Steinbeis Transferzentrum für Elektromagnetische Verträglichkeit, Kommunikationstechnik und Hochfrequenztechnik (EKHO), bei dem ich seit Dezember 2015 parallel zu meiner Promotion ebenfalls tätig war.

Ein Dankeschön an Herrn Rusnak und Herrn Catrinescu von der Firma Embedded Wireless GmbH für die sehr gute Zusammenarbeit beim Forschungsprojekt CHAMELEON und die Unterstützung bei allen Fragen zu Ihrer Hardware.

Vielen Dank auch an Tom Henderson und das ns-3 Team für den Support bei Fragen zum Netzwerksimulator.

Zuletzt möchte ich mich bei meiner Familie bedanken, welche mir immer großes Vertrauen entgegengebracht sowie die Unterstützung und Freiheit gegeben hat, meinen eigenen Weg zu gehen. Da meine Mutter bereits während dem Grundstudium und meine Oma während der Promotion verstorben sind, sei diese Arbeit diesen beiden wunderbaren Personen gewidmet.

Ulm im Januar 2019   Christopher Hepner

Kurzfassung

Nach dem Auftreten einer Katastrophe (z. B. eines Erdbebens oder eines Tsunamis) ist eine funktionierende Kommunikationsinfrastruktur eines der Hauptbedürfnisse von Rettungsteams und freiwilligen Helfern, auch bereits in den ersten Stunden nach dem Auftreten des katastrophalen Ereignisses. Das Disaster Recovery-System (DRS), welches in dieser Arbeit vorgeschlagen wird, basiert auf einem IEEE 802.11s Wireless Mesh Network, welches sich aus unbeschädigten, wireless-mesh-fähigen, batteriebetriebenen Geräten zusammensetzt, die im Katastrophengebiet noch verfügbar sind. Wichtige Überlegungen hinsichtlich der Performance eines DRS sind die erforderliche Knotendichte, die Abdeckung, die durch eine gegebene Knotendichte erreicht werden kann, die Skalierbarkeit sowie die Lebensdauer des Systems. Zur Erhöhung der Lebensdauer, welche eine der Schlüsselherausforderungen für das vorgeschlagene batteriebetriebene DRS darstellt, wird ein verteilter Algorithmus vorgeschlagen. Der Ansatz ist hierbei, nicht benötigte Knoten abzuschalten und für einen späteren Zeitpunkt der Verwendung aufzubewahren, während die Netzwerkkonnektivität erhalten bleibt.

Dieser Algorithmus basiert auf einem neuartigen Ansatz, bei dem die Entscheidung, wann ein Knoten seinen Zustand vom Ruhezustand zum aktiven Zustand oder umgekehrt ändert, auf der Anzahl der vorhandenen Nachbarknoten des entsprechenden Knotens basiert, was eine einfache Implementierung auf den einzelnen Mesh-Knoten ermöglicht. Die erreichbare Lebensdauerverlängerung wird zunächst durch eine Monte-Carlo-Simulation untersucht. Darauf aufbauend wird die mit dem vorgeschlagenen Lifetime Enhancement-Algorithmus für Disaster Recovery-Systeme (LEA-DRS) erreichbare Netzwerkperformance durch eine Netzwerksimulation mittels des Open-Source Network-Simulators ns-3 validiert.

In Netzwerksimulationen ist eine korrekte Repräsentation, insbesondere von Schicht 1 (Physical Layer) und Schicht 2 (Data Link Layer) notwendig, um zuverlässige Ergebnisse zu erzielen, die mit der realen Hardwareperformance vergleichbar sind. Das Wi-Fi-MAC-Modell des Netzwerksimulators ns-3 für Wireless Mesh Networks wird in dieser Arbeit zunächst validiert und es werden geeignete Korrekturmaßnahmen für die aufgetretenen Ungenauigkeiten der implementierten Modelle vorgeschlagen.

Des Weiteren werden Verzögerungszeiten, die in jeder Wireless Mesh-Station aufgrund von Software-Ausführungszeiten und des Scheduling-Mechanismus von Tasks in der CPU auftreten, im Modell hinzugefügt. Diese werden durch geeignete Hardwareaufbauten und Messungen von mehreren WLAN-Modulen bestimmt und im Netzwerksimulator realisiert, was schließlich zu Simulationsergebnissen führt, die eine möglichst große Vergleichbarkeit mit in Hardware realisierten Netzwerken erreichen.

Die erreichbare Netzwerkperformance wird abschließend validiert, indem ein geeignetes Physical Layer-Modell einschließlich der Verzögerungszeiten und zusätzlicher Hardware-Parameter verwendet wird. Die erzielten Ergebnisse zeigen, dass eine Kommunikation per VoIP in einem Katastrophengebiet mit angemessener Performance und Lebensdauer auf Basis des entwickelten Systems möglich ist.

Abstract

After the occurrence of a disaster (e. g. an earthquake or tsunami) one of the main needs for the rescue teams and volunteer helpers is a functional communication infrastructure even during the first hours. The disaster recovery system (DRS) which is proposed in this work is based on an IEEE 802.11s wireless mesh network which is set up by non-damaged, legacy, mesh capable and battery powered devices still available in the disaster region. Key considerations concerning the performance of a DRS are the required node density, the coverage that can be achieved by a given node density, scalability as well as the lifetime of the system. For lifetime enhancement, which is a key challenge for the proposed DRS based on battery powered devices, a distributed algorithm is proposed which enhances the lifetime of such a system by an approach which allows to shut down non-necessary nodes and to keep them for a later usage while still keeping network connectivity.

The proposed algorithm is based on a novel approach whereby the decision when a node changes its state from sleep to awake or vice-versa is based on the number of neighbors of the corresponding node, allowing for a simple implementation on the individual mesh nodes. The achievable lifetime enhancement first is evaluated through a Monte-Carlo simulation. In a second step, the network performance achievable with the proposed lifetime enhancement algorithm for disaster recovery systems (LEA-DRS) is validated by a network simulation in the open-source network simulator ns-3.

In network simulations a correct representation of the physical layer and the MAC layer is essential in order to achieve reliable results which are comparable to real hardware performance. To this end, the Wi-Fi MAC model of the network simulator ns-3 for wireless mesh networks is validated in this work and corrections for the inaccuracies which are present in the models are proposed.

Furthermore, processing delays which are added in each wireless mesh station due to software execution time and the scheduling mechanism of tasks in the CPU are estimated by appropriate testbed measurements of WLAN modules and included in the network simulator leading to more appropriate simulation results.

The achievable network performance is finally validated by using an appropriate physical layer model including processing delays and additional hardware parameters. The results show that a communication by VoIP will be possible in a disaster area to a reasonable extent and over sufficient lifetime when using the DRS proposed in this work.

Contents

Chapter 1

Introduction

Contents

1.1 Motivation

After the occurrence of a disaster such as an earthquake, tsunami, hurricane or flood most of the communication infrastructure is usually destroyed or not working due to the loss of power. However, providing a suitable communication network in disaster areas – called a Disaster Recovery System (DRS) throughout this work – as quickly as possible is one of the most essential needs for rescue managers or people in need in order to organize help. This importance is growing also in view of the increasing number of natural disasters over the last 40 years. Figure 1.1 shows the accumulated increase of geophysical (e. g. earthquake, mass movement, volcanic activity), hydrological (e. g. flood, landslide) and meteorological (e. g. storm) disasters sorted by continents since 1975 according to the International Disaster Database of the Centre for Research on the Epidemiology of Disasters (CRED) [1]. Some of the biggest natural disasters to be mentioned in the last years are the floodings during 2010/2011 in China with up to 134 million affected people, the flooding during 2010 in Pakistan with over 20 million people affected and 1985 deaths, Typhoon Haiyan in 2013 which affected over 16 million people in the Philippines and caused 7354 deaths, further the Haiti 2010 earthquake which affected about

3.7 million people and caused 222.570 deaths, the great earthquake in Japan during 2011 which affected 368.820 people and caused 19.846 deaths and the Nepal 2015 earthquake which affected over 5 million people and caused 8831 deaths. Recently hurricanes passed in late 2017, e. g. Harvey, Irma and Maria in central America and earthquakes, e. g. in Mexico strengthen the need for appropriate disaster recovery systems.

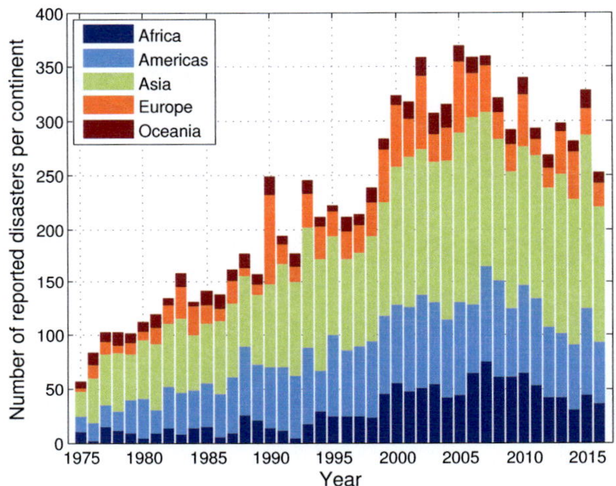

Figure 1.1: Total number of reported natural disasters between 1975 and 2016 [1]

While nowadays help usually arrives within a few days it still happens that some regions are isolated for weeks after natural disasters. However, the great earthquake in East Japan (2011) or the Pakistan flood (2010) showed that the rescue teams are composed only of very few trained professionals, but of hundreds of disorganized volunteers during the first hours. Therefore it has been proposed, e. g. in [2], [3], [4], [5], [6], [7] and [8], to use the infrastructure which is still working or to use devices carried by the rescue teams and volunteers themselves or a combination of both to set up a communication system.

Because of the so called "golden 72 hours" [3] – which are the most important ones for successful rescue operations – a communication system for the rescue volunteers and rescue teams should instantaneously be available without any deployment. Almost immediate availability and a sufficient lifetime of a communication infrastructure after occurrence of a disaster can thus be considered as key elements of successful operations of search and rescue and is the key subject of this research work.

Additionally, during the last years internet and social media became more and more important in the aftermath of natural disasters. Volunteers at the East Japan great earthquake (2011) used the internet to organize their activities [9]. A study of NetHope [10] shows that spontaneous local volunteers were using internet-based social media on mobile devices to reach out for help during the 2010 Pakistan flood. Humanity Road Analytics details that the most commonly searched terms used by the public when viewing their website were the need for "shelters" followed by "find hospital" [10]. Google engineers built the "Google Person Finder" in response to the January 2010 Haiti earthquake in order to help those affected by the earthquake to connect with their relatives [11]. It is an open source web application that allows individuals to post and search for the status of relatives or friends affected by a disaster. Since then the software was used e.g. in the February 2010 Chile earthquake, July 2010 Pakistan floods (where it didn't work well because people in the affected area had no internet access [12]), February 2011 Christchurch earthquake, March 2011 Tohoku earthquake and tsunami, October 2011 Van earthquake, April 2013 Ya'an earthquake, October 2013 cyclone Phailin, November 2013 typhoon Haiyan and April 2015 Nepal earthquake. Recently Google has also activated the Person Finder service in the aftermath of hurricane Maria in 2017.

Recently the Federal Office of Civil Protection and Disaster Assistance – Bundesamt für Bevölkerungsschutz und Katastrophenhilfe (BBK) – in Germany has also published a guide for emergency preparedness and correct action in emergency situations. The document focuses on personal preparedness of each individual ([13], 2016). In general it is said: Germany is well prepared. A number of organizations such as fire brigades, the police and rescue services are there for everyday aid. However, it is also stated that even the best assistance is not always on the spot immediately and in the event of a large-scale and very serious disaster the rescue workers cannot be everywhere. Everyone is encouraged to be able to help themselves. The BBK has also developed an Emergency Information and Warning App called NINA (Notfall-Informations- und Nachrichten-App). The App is intended to warn users about emergencies and hazards all over Germany, such as severe weather, floods and other relevant events. The push functionality calls the attention of the users always to current threats. However, it has to be emphasized that during an electric power breakdown, mobile phones will only continue to work for a short time and these services are no longer available [13].

In 2015 the BKK has started a collaborative research project with the Technical University of Darmstadt and the University of Kassel, called "smarter". "smarter" stands for Smartphone-based Communication Networks for Emergency Response and allows, by the help of a smart phone App that was developed during the project, the direct communication from smart phone to smart phone via a Wi-Fi based ad-hoc network. The research partners thereby examined different areas covering the technical feasibility, the legal framework or the behavior and needs of people in crises and disasters. In the smarter project the BBK looked at the behavior of the population in crises and catastrophes. To this end a field test was carried out during the project in order to examine how the participants used the App and which functions they considered to be of particular importance. The preliminary evaluation of the data in [14] showed that it was most important for the participants to exchange messages, to call for help and to receive valuable information. [14]

1.2 Objective

For the immediate needs of communication after the occurrence of a disaster, approaches that use the infrastructure which is still available ([3], [5], [8]) seem to be most suitable. In this work a new approach for this kind of DRS is proposed based on the mesh network amendment for Wireless Local Area Networks (WLANs) IEEE[1] 802.11s [15]. The focus thereby is on enhancing the lifetime of the DRS in order to bridge the time until professional rescue teams with their communication equipment will arrive while still keeping a network connectivity that provides essential communication needs, especially Voice over IP (VoIP) and low-rate data services. It is assumed that WLAN interfaces will be available in many systems in the future such as smart phones, laptops, cars and home appliances where most of them are battery powered or at least have an auxiliary battery. All these WLAN devices in the disaster area are assumed to have mesh functionalities and therefore are called Mesh Stations (MSTAs). Because of the number of devices available in disaster areas and the ease of networking without the need of additional infrastructure a wireless mesh network built from the devices that are still functional after the occurrence of a disaster may provide a very attractive way to set up a communication network in a disaster area already during the first hours.

Currently, WLAN based mesh networks have their primary usage scenario in the domain of public wireless access where the wireless mesh network provides a flexible backhaul for WLAN access points, which are distributed throughout cities or university and company campuses. In the future, however, it seems quite probable that the mesh network capability might be available in a high number of WLAN devices, e. g. through its availability in open source implementations such as OpenWrt [16]. In order to protect owners of the MSTAs from illegal usage and law infringements by unwanted establishment of a DRS a mechanism such as an emergency mode for the MSTAs ([5], [17]) has to be used. During an emergency situation the wireless stations will leave their usual mode of operation and will switch into the emergency mode and thus can be used by the rescue teams or people in need. Let us remark that the activation of this mode is out of the scope of this work.

Objectives of this work are to propose a Wireless Mesh Network (WMN) based DRS using legacy IEEE 802.11s, to develop a suitable algorithm for lifetime enhancement of the DRS and to investigate the achievable network performance of such a DRS.
The system coverage of the DRS will be investigated by system simulations carried out in Matlab and the achievable network performance will be analyzed by network simulations with the network simulator ns-3. In order to get more reliable results from the network simulations which additionally account for constraints introduced by the underlying hardware, the ns-3 network simulator is further extended by including hardware parameters which are obtained by small testbed measurements in a laboratory environment.

---

[1]Institute of Electrical and Electronics Engineers (IEEE).

Let us make the proposed DRS more explicit by giving a simple example. Figure 1.2 shows a sketch of a disaster recovery situation after an earthquake with connected Mesh Stations in home appliances (e. g. Access Points (APs) with Mesh capability) and cars. All MSTAs which are within the transmission range of each other establish peer links in between them and thus allow the routing of data throughout the mesh network formed by them. The communication is routed from handheld devices (e. g. smart phones, laptops) of the rescuers through an essentially static network formed by the remaining devices (e. g. battery powered MSTAs in cars, Access Points with mesh functionality and smart home devices) that are still working. One or more MSTAs in the mesh network may also act as a gateway to another network. Figure 1.2 shows the communication between two users by their smart phones over the mesh network as an example.

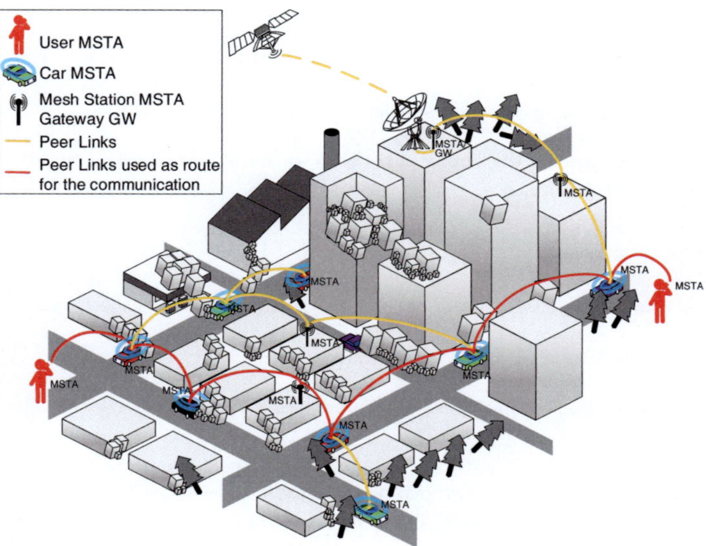

Figure 1.2: Sketch of Disaster Recovery System (DRS) after an earthquake built from connected Mesh Stations (some of them with gateway capability to other networks) in home appliances and cars.

Smart phones and laptops are assumed to have a relatively low battery lifetime between zero and 24 hours. This also holds for currently available battery powered travel routers. On the other hand, WLAN interfaces available in cars or in some other types of battery powered electrical devices may have much longer battery lifetimes of several days because of the powerful batteries in these devices. It is assumed that the number of such devices will be strongly increasing throughout the coming years by the idea of the "internet of things" and the "connected car". A WMN serving as DRS could thus be set up by the MSTAs available in numerous devices which are not destroyed. Volunteers as well as victims using their own smart phones or laptops may use this WMN and communicate within the covered area. Several services such as voice over IP, push to talk, video streaming or a map of the disaster area can be kept in mind as useful services for the rescue teams. Even an injured person who is located under rubble could set up an emergency call over the WMN.

The solution proposed in this work covers the following requirements that a disaster recovery system will have to fulfill:

1. System requirements:

   - Instantaneous availability of the system within the first hours after the occurrence of a disaster.

   - Formation of the system by standard user equipment.

   - Access to the system with standard user equipment.

   - Enhancement of the lifetime of the DRS while still providing sufficient coverage and network performance.

2. Service requirements:

   - Possible communication of rescuers and victims by a VoIP connection.

   - Possible exchange of text or voice messages.

   - Internet connection, if MSTAs with network access and gateway capability are still available.

These requirements have to be provided over a maximum period of time. Therefore a distributed algorithm is proposed which enhances the lifetime of such a system by an approach which allows to shut down non necessary nodes in order to save their battery capacity for later usage while still keeping network connectivity.
The difficulty thereby is to validate the functionality of such an algorithm with a huge number of MSTAs which are distributed in the disaster area. Testbed measurements of such a network with a large number of MSTAs would be time-consuming, expensive and rather difficult to set up. Network simulation on the other hand allows the relatively fast assessment even of complex scenarios with reproducible results and relatively simple change of settings, protocols and environment. However, it may provide unrealistic results if hardware aspects are not properly considered. In order to overcome these challenges an incremental approach, by successively increasing the complexity of the model and successively taking hardware influences into account is proposed, finally leading to a realistic modeling of the system, covering the most relevant parameters.

Figure 1.3 summarizes the methodology developed in this thesis and how it is used to develop a DRS which is capable of achieving the proposed requirements. Further the tool chain, which is used in this work is briefly outlined in Fig. 1.3.

First, the question of how many MSTAs are needed to achieve a connected network in a disaster area of given size is investigated based on state-of-the-art analytical calculations assuming toroidal boundary conditions and thus allowing to neglect boundary effects (Fig. 1.3 a).

In order to additionally account for boundary effects the achievable connectivity will be further calculated by Monte Carlo simulations carried out in Matlab for a dedicated amount of nodes N within finite rectangular areas. Of specific interest for what follows is the covered area which can be reached by the largest subset of connected nodes $n_1, \ldots, n_{N_c(t=0)}$ and its associated probability which will be called the coverage of the DRS. These quantities will be evaluated by Monte Carlo simulations on system level carried out in Matlab (Fig. 1.3 b).

The achievable lifetime of the network can be calculated by extending the simulation approach in Matlab to cover the finite lifetime of the individual nodes (Fig. 1.3 c). This is achieved by adding a certain lifetime to each node and calculating the coverage of the remaining nodes $n_1, \ldots, n_{N_a(t_j)}$ for dedicated time instances $t_1, \cdots, t_m$. Coverage thereby again is characterized by the area covered by the largest subset of connected nodes $n_1, \ldots, n_{N_{a,c}(t_j)}$ still available at a given time instance t_j and its associated probability.

In order to enhance the lifetime of the system a distributed algorithm will be developed. This algorithm denoted as Lifetime Enhancement Algorithm (LEA) allows to shut down redundant nodes and reactivate them later when needed (Fig. 1.3 d). The algorithm works on basis of the IEEE 802.11s Mesh Peering Management (MPM) protocol by developing an appropriate metric for the decision of when to shut down and when to reactivate an individual node. Again the coverage of the remaining active nodes which are still working and not shut down $n_1, \ldots, n_{N_a(t_j)}$ for dedicated time instances $t_1, \cdots, t_m$ can be calculated demonstrating the achievable extension of system lifetime at given coverage requirements.

In order to evaluate the network performance that can be achieved by the DRS, the protocol stack for the WMN has to be taken into account. The statistical network performance thereby is evaluated on basis of VoIP connections between two randomly placed nodes (e. g. two rescuers) using network simulations within the network simulator ns-3. As a first step (Fig. 1.3 e) the network performance at a given time instance t_j is investigated by using the distribution and positions of the active nodes $n_1, \ldots, n_{N_a(t_j)}$ as calculated by the implementation of the Lifetime Enhancement Algorithm (LEA) in the Matlab based system simulation. This type of network simulation in ns-3, which does not provide for an implementation of LEA within the network simulator itself we will call the so called static simulation. The static simulation thereby allows to analyze the achievable network performance of LEA without the additional influences of the protocol stack and physical layer during the execution of the LEA itself. Thus the static network simulation provides information on the network performance of the thinned out network that is established by LEA under the idealized assumption that LEA itself is not affected by the network protocol and wireless channel performance.

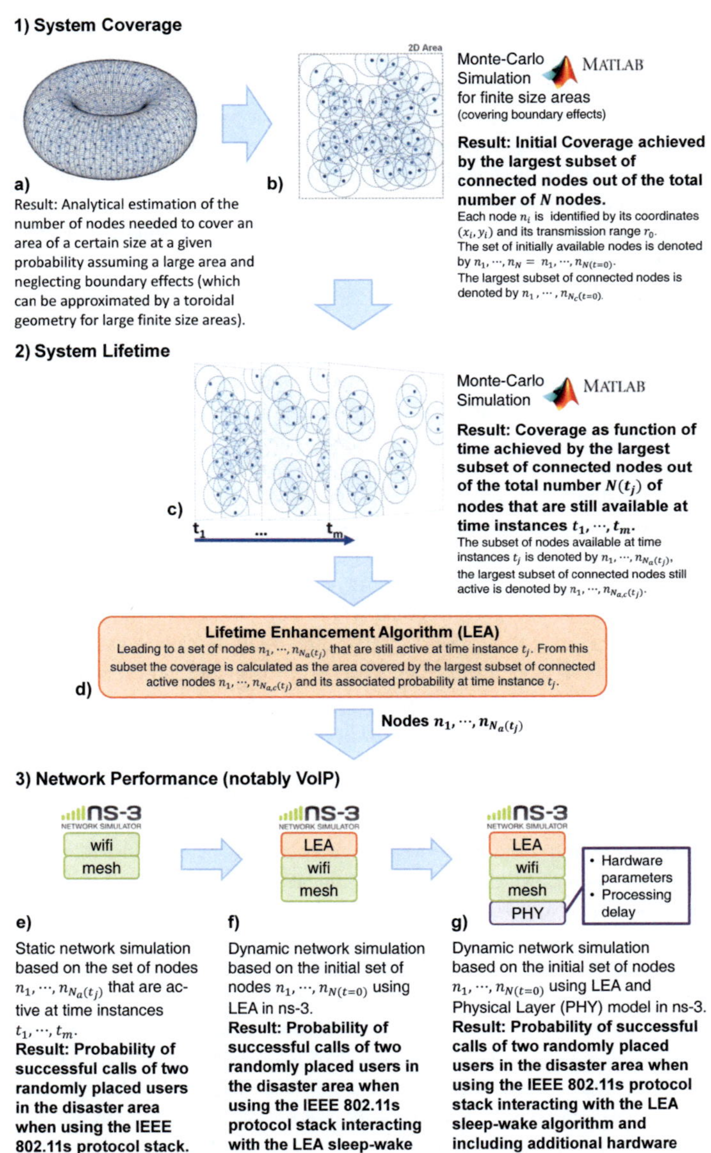

Figure 1.3: Overview on the methodology and toolchain used throughout the thesis in order to achieve the objectives of this work.

The key result of the static network simulation will be the probability for successful calls that can be achieved between two randomly placed users and is obtained by carrying out a sufficiently large amount of Monte Carlo simulation runs.

Second (Fig. 1.3 f), the DRS network performance will be evaluated by using a full implementation of the distributed algorithm within the network simulation itself which is based on alternating sleep and awake states for each individual node. In that way the algorithm can be analyzed in conjunction with the whole protocol stack. This type of simulation we will call the dynamic network simulation. Again the key result is the probability of successful VoIP calls between two users randomly placed in the disaster area and again it is obtained by carrying out a sufficiently large amount of Monte Carlo simulation runs.

Finally (Fig. 1.3 g), the network performance will be analyzed under the additional influence of an appropriate physical layer model which is based on explicit hardware parameters obtained from test network measurements with appropriate hardware modules. Key hardware parameters to be determined thereby are processing delays and packet error rates. Both parameters are investigated by measurement setups requiring only a small number of MSTAs, e.g. in small testbed arrangements. Using appropriate testbed arrangements it will be possible to measure the delay which is added by each hop in a multihop scenario. In that way the processing delays which are added in each MSTA due to the software execution time and scheduling mechanism of tasks in the CPU can be reliably estimated. In the framework of appropriate testbed setups the error rate performance of IEEE 802.11s hardware modules is evaluated and correlated with the physical layer model employed in ns-3. The ns-3 physical layer model is based on the calculation of Bit Error Rates (BERs) taking into account the forward error correction present in IEEE 802.11 and will be compared with hardware measurements.

The final results including these additional hardware parameters (i.e. an appropriate physical layer model based on the BER-model, hardware parameters as well as an appropriate processing delay model) will indicate the feasibility and possible lifetime of a IEEE 802.11s based DRS and the achievable network performance assuming a VoIP connection of two rescuers.

Part of this work has been published throughout the last years by the author. The influence of processing delays on the VoIP performance for IEEE 802.11s multihop wireless mesh networks and further the comparison of ns-3 network simulations with hardware measurements has been published in [18]. The Lifetime enhancement of Disaster Recovery Systems based on IEEE 802.11s Wireless Mesh Networks and further the enhancement by using a sleep-wake algorithm with a minimum number of neighbors has been published in [19] and [20]. The analysis of the ns-3 physical layer abstraction for WLAN systems and evaluation of its influences on network simulation results has been published in [21] and [22]. A validation of the ns-3 802.11s model finally has been published in [23].

1.3 Outline

This thesis is structured as follows:

Before developing the DRS system design in Part I of this thesis, some fundamental considerations are presented in Chapters 2 and 3.

Chapter 2 thereby gives an overview on the state-of-the-art of research on Disaster Recovery Systems and the IEEE 802.11 Wireless LAN Medium Access Control (MAC) and Physical Layer (PHY) specifications.

Chapter 3 provides evaluation methods for coverage, connectivity and network performance of Wireless Mesh Networks. Specifically an appropriate method for the calculation of the coverage of a DRS is developed.

For a successful usage of the ns-3 network simulator in order to evaluate the performance of IEEE 802.11s WMNs based DRS a number of modifications and amendments had to be developed throughout this thesis. The second part of Chapter 3 is dedicated to a presentation of these modifications including a validation of the resulting MAC and PHY implementations in ns-3 based on both analytical calculations and simulation.

The thesis is then organized in two main parts. Part I discusses the system design. First the system concept of the DRS based on the requirements is shown in Chapter 4. The system coverage will be analyzed by analytical and simulative approaches and the system lifetime will be investigated based on Monte Carlo simulations.

Chapter 5 then presents the algorithm for the lifetime enhancement (LEA) of the DRS, carries out a proof of concept and finally develops the distributed sleep-wake algorithm acting on each individual MSTA.

Part II presents the system implementation and validation. First the network performance of the distributed lifetime enhancement algorithm is evaluated in Chapter 6.

Chapter 7 details the additional hardware requirements on the air interface. In particular the rerouting performance evaluation and the influence of processing delays on the VoIP-performance will be shown.

Chapter 8 shows the final system performance of the DRS including additional hardware parameters which will indicate the feasibility and possible lifetime of a IEEE 802.11s based DRS as well as the achievable network performance assuming a VoIP connection between two rescuers.

Chapter 9 concludes the thesis work and presents suggestions for future work.

Chapter 2

State of the Art of Disaster Recovery Systems and Basic System Considerations

Contents

2.1 Disaster Recovery Systems using Wireless Mesh Networks

Generally, under the expression "Disaster Recovery System" all aspects of disaster recovery are meant, e. g. the recovery or emergency restoration of water- or power-supplies, road infrastructure, communication and all other personal needs after a disaster has happened. However, one of the main needs for rescue teams and volunteer helpers after the occurrence of a natural disaster (e. g. an earthquake), is a functional communication infrastructure especially during the first 72 hours. During these first hours victims have the highest surviving rate after geological disasters. This section therefore focusses only on the state-of-the-art of the recovery of communication systems which are the focus of this research. Thus the concept of a Disaster

Recovery System (DRS) is used throughout this text to exclusively denote a system which is able to restore a certain amount of communication infrastructure. Thereby it has to be distinguished between governmental Emergency- and Rescue Services (ERS) and other systems for voluntary helpers and people in need which are denoted as Public Disaster Recovery Systems in this text.

ERS are organizations which ensure public safety and health, e.g. law enforcement, fire departments and emergency medical services. Communication systems for governmental services are generally installed before an incident and require heavy network planning and organization. Of course in this case it has to be distinguished between industrialized and developing countries where in developing countries a comprehensive implementation of area-wide systems cannot be expected. Also in such disaster situations usually professional search and rescue teams from other countries assist the locals and the foreign rescue teams normally carry their own communication equipment along with themselves. Interoperability of those systems of organizations from different countries thereby is an important issue. In the 2010 Haiti earthquake, e.g. first rescue teams from other countries arrived within the first day of the incident. During the Great East Japan Earthquake (2011) 163 countries and regions as well as 43 international organizations had offered aid and relief. Emergency assistance squads, medical teams and reconstruction teams had been dispatched from 24 countries and regions along with expert teams from five international organizations [24].

Public disaster recovery systems can be classified into two principal categories. First, systems which are established before a disaster has happened (so-called pre-DRS) and second, systems which are established after an incident (so-called post-DRS). For both types different solutions based on multiple technologies and protocols are proposed. The advantages of pre-DRS are a fast establishment and planed system coverage, however, with the drawback of high expenses which make them almost only possible in developed countries[1]. Post-DRS form the majority of the currently described systems for public disaster recovery which are mostly using systems still available in the disaster region. This provides for cheap solutions without the need of system planning. The disadvantage, however, is that system coverage cannot be guaranteed.

2.1.1 Governmental Emergency and Rescue Services

Traditional communication systems of ERS rely on Land Mobile Radio (LMR). Those systems provide analog voice communication for closed user groups over dedicated Ultra-High Frequency (UHF) or Very High Frequency (VHF) radio-frequency bands [25]. Modern systems are digital with limited data capabilities. The two most relevant standards to be mentioned in this context are the Association of Public Safety Communications Official (APCO) Project 25 (P25),

---

[1]It should be noted that governmental Emergency- and Rescue Services also belong to the category of pre-DRS.

standardized by the Telecommunications Industry Association (TIA), and TErrestrial Trunked RAdio (TETRA) [26], developed by the European Telecommunications Standards Institute (ETSI) [25]. Beside North America, APCO-P25 is used in Australia, India, Russia, Singapur and some countries in South America. TETRA is the de-facto standard in Europe. The Asia Pacific region shows heterogeneous deployments, where both TETRA and APCO-P25 technologies exist beside each other.

Both types of ERS are not immune to failure when a disaster has happened which will be further described exemplary by the ERS in Germany.

A case study of the Federal Office of Civil Protection and Disaster Assistance – Bundesamt für Bevölkerungsschutz und Katastrophenhilfe (BBK) – in Germany during 2011 details the crisis management in the case of a large interruption of the power supply using the example of the constituent state Baden-Württemberg[2] [27]. Causes of power failures described in [27] are natural disasters (e. g. storm, thunderstorm, flood, earthquake), technical and human failure (e. g. malfunctions due to ageing, design defects, poor maintenance), deliberate acts (e. g. terrorist attack, sabotage) or network overloads and system failure (e. g. instability of frequency or voltage as a result of the protective switch-off of equipment). It is assumed that all causes tend to increase in number either because of climatic changes, higher complexity of technical systems, changes in political situation or due to increased load flows in the liberalized energy market. Even though an earthquake is not as likely as in other countries, the probability of a blackout even in Germany seems to be increasing due to the mentioned reasons. Table 2.1 shows the impact of power failures on the information and communication technology according to [27], differentiating between three scenarios. Scenario A with power failures with a duration of less than eight hours, Scenario B with power failures from eight up to 24 hours and Scenario C with power failures greater than 24 hours.

Mobile communications (e. g. GSM, UMTS, LTE, WiMAX)[3] and fixed communication will last only few hours or immediately fail depending on their backup with uninterruptible power supplies (UPS). Also internet and data networks will fail depending on the UPS of routers or servers and the state of battery of end-user-equipment, e. g. laptops.

During the last years the outdated analog radio network has been replaced by the new BOS[4] digital radio network based on the TETRA standard which builds a single network for all security authorities and organizations in Germany while the migration is still ongoing.

As of December 2016, 4521 base stations are in operation. This means 99% of the area of Germany is covered. One of the main reasons for the system change, beside a very high and reliable availability and improved speech quality, was the security against interception of radio communications. The German BOS digital radio network is now the biggest TETRA-based radio network worldwide with currently 692.000 users [28]. According to the Federal Agency for

---

[2]Germany consists of 16 constituent states (Länder, singular - Land); Germany is a federal republic and each constituent state has its own parliament and government, and a high degree of autonomy.

[3]Global System for Mobile communication (GSM), Universal Mobile Telecommunications System (UMTS), Long Term Evolution (LTE), Worldwide Interoperability for Microwave Access (WiMAX).

[4]Behörden und Organisationen mit Sicherheitsaufgaben (BOS) – Authorities and organizations with security tasks, e. g. ERS.

Table 2.1: Impact of power failures on information and communication technology. [27]

Field	Scenario A (< 8h)	Scenario B (8-24h)	Scenario C (> 24h)
Mobile communications	– Immediate outage of unsecured base stations – Failure of UPS-assured base stations (2h) – Failure of Base Station Controller (BSC)(about 4-6h) – Network congestion	– Failure of mobile phones (depending on the charge state of batteries) – Failure of emergency power supplied Base Stations	– Fuel shortage for Emergency Power Supplies (EPS) – Failure of Mobile Switching Center (MSC) (about 4 days) – Failure of mobile communication equipment (about 4-6 days without calls)
Fixed communications	– Failure of cordless phones (without battery in base station) – Failure of telephones without emergency mode – Failure of DSL modem / router	– Failure of cordless phones (with battery in base station, depending on charge level) – Failure of ISDN telephones with emergency mode – Partial failures in the network	– Failure of cordless phones (with battery in base station, depending on charge level) – Failure of Emergency Supply of central switching centres (3-4 days) – Fuel shortage for Emergency Power Supplies
Internet and Data Networks	– Failure of routers, switches – Failure of modems – Failure of not UPS-secured servers – Failure of PCs and laptops (2-5h)	– Failure of laptops	– Fuel shortage for Emergency Power Supplies – Failure of Emergency Power Supplies of data centers (about 1 week)
Emergency Services	– Failure of relay stations (analog radio) (UPS 4-8h) – Failure of Base Stations (digital radio) (about 2h, battery operation)	– Failure of relay stations (dependent on emergency supplies)	– Fuel shortage for Emergency Power Supplies – Failure of mobile relay stations (dependent on emergency supplies)

Public Safety Digital Radio – Bundesanstalt für Digitalfunk der Behörden und Organisationen mit Sicherheitsaufgaben (BDBOS) – a minimum of two hours UPS must be guaranteed. The respective federal state, however, determines the way in which the supply of the base station is ensured beyond the UPS system and realizes the solution on its own responsibility. The emergency power supply (EPS) for the TETRA switching centers is provided via redundant EPS, which ensure a supply for the duration of at least 72 hours with the fuel stored on site [28]. It therefore can be concluded, that even though a single network for all security authorities and organizations was established a working network is only guaranteed two hours after an incident and the actual working time may differ from federal state to state.

A review of the Great East Japan Earthquake [24] shows also the vulnerability of the governmental emergency- and rescue networks after a natural disaster. It is stated that during the response stage satellite phones played a crucial role in the emergency communication for local governments and rescue organizations. The reason is that the government radio communication infrastructure was seriously damaged and emergency- and rescue services, e. g. fire departments in devastated areas had lost their radio equipment or base of communications.

Further an immense damage and congestion of telephone infrastructure, including 1.9 million fixed-line services and 29.000 mobile-phone base stations was observed. For people in need social media played a crucial role and was extensively used for search and rescue of those who still had connection to the internet. These include social networks such as Twitter or Facebook to connect with other people as well as the Google Person Finder which allows people to enter an inquiry about a missing person. [24]

The case study of the Federal Office of Civil Protection and Disaster Assistance in Germany [27] and the review of the Great East Japan Earthquake [24] both show that even governmental emergency- and rescue networks are vulnerable and not sufficient to ensure the communication in the aftermath of a disaster. It is also clear, that for the disaster recovery of communication networks for the public other solutions have to be envisaged.

2.1.2 Public Disaster Recovery Systems and Related Research

The majority of currently proposed public DRS belong to the group of so-called post-DRS which are mostly using systems still available in the disaster region. The following section gives an overview on currently proposed systems in research literature and discusses these approaches in view of the objective of this thesis.

Focus thereby will be twofold: On one hand research on suitable communication network solutions will be discussed, on the other hand possible applications that can provide for the required services in a disaster situation will be discussed, independently of the underlying network solution.

Regarding suitable communication networks [2], [3], [4] and [5] are examples of current research based on WLAN mesh and ad-hoc networks. In [2] a DRS is proposed were users can get access to the network using their own smart phones or PCs with IEEE 802.11 WLAN by using a mesh backbone network with mesh nodes installed in street-side LED town lights or even Unmanned Aerial Systems (UAS). In [3] it is proposed to use Wi-Fi capable laptops or other handheld devices carried by the volunteers themselves to set up a Mobile Ad-hoc Network (MANET). In [4] a solution is proposed, which connects remaining WLAN-routers together with a Movable and Deployable Resource Unit (MDRU) serving as a base station and providing internet connectivity. [5] proposes an on-the-fly establishment of multihop wireless access networks (OEMAN) for disaster response by extending the internet connectivity from surviving WLAN access points to disaster victims using their own mobile devices.
Some research papers such as [6], [7] and [8] propose multi-tier architectures which may consist of several air interfaces, e. g. Satellite, WiMAX, LTE, UMTS, GSM, WLAN and which usually require system concepts across several network layers. As an example [6] exploits a solution which allows the delivery of mission-critical multimedia data between rescue teams and their

headquarters over extremely long distances using a combination of wireless network technologies (namely Wi-Fi, WiMAX and GEO Satellite) to meet the requirements of a disaster rescue communication scenario. However, it has to be recognized that all these solutions which rely on several network layers have a high complexity and therefore are subject to a relatively high risk of outage. [7] provides a solution which allows free voice calling and text messaging within a disaster area, and enables users of unmodified GSM handsets to communicate with the outside world using the Skype VoIP network. [8] proposes a new mobile network architecture (including 3GPP LTE and Wi-Fi radio access technologies) based on distributed controls and ad-hoc configurations. In this architecture the locally available resources such as frequency spectrum and lively nodes are used to reconfigure the network and provide emergency services. The mobile ad-hoc networking functionality which is proposed in [8] allows users normally covered by access points or base stations that stopped to operate after a disaster to be able to communicate using intermediate nodes connected to access points and base stations that are still working. For this purpose smart phones capable of sharing Internet connections are used in combination with mesh topologies where the explicit approach is not further described.

It should also be noted that the solutions provided by [6], [7] and [8] are dependent on a minimum of infrastructure still working.

All the proposals (e. g. [4], [6], [7]) relying on infrastructure which has to be taken to the disaster area and set up by the rescue teams themselves have the disadvantage of a considerable time of deployment. Because of the so-called "golden 72 hours" [3] – which are the most important for successful rescue operations – a communication system for the rescue volunteers and rescue teams should instantaneously be available without any deployment.

On the other hand, systems like [2] belonging to the group of pre-DRS with mesh nodes installed in street-side LED town lights have to be installed prior to the occurrence of a disaster and need a considerable amount of network stations and network planning.

Almost all of the previously mentioned systems are research proposals and have not been used in disaster areas so far. The use of the Movable and Deployable ICT Resource Unit (MDRU) described in [4] has been tested in a feasibility study on restoring the telecommunication, information and communication technology in San Remigio municipality on Cebu Island in the Philippines which sustained enormous damage from typhoon Haiyan in November 2013. The study was carried out by the NTT Network Innovation Laboratories in cooperation with the Ministry of Internal Affairs and Communications of Japan, the Department of Science and Technology of the Philippines and the International Telecommunication Union (ITU) in a test setup from May 2014 till September 2015. The MDRU was installed in the San Remigio Municipal Hall and the wireless equipment was installed in a national high school (about 400 m away from the hall), where an evacuation center had been set up. A communication link was provided by a point-to-point wireless link between the municipal hall and the high school. Two wide area Wi-Fi networks, one at the municipal hall as well as another one at the high school were established by several Access Points (APs) and a 24-GHz Fixed Wireless Access (FWA) link between the two buildings connecting them. It was confirmed through the feasibility study that the MDRU operated effectively in the environment of the test site at the Philippines. As a consequence NTT and NTT Com are working to promote broad use of the MDRU throughout the world for disaster relief after large-scale disasters [29].

Another system to be mentioned which focuses not on disaster recovery but on the in-vehicle emergency call (e-Call) after a car accident is the eCall system which has been recently adopted by the European Parliament and standardized by ETSI. The eCall regulation requires all new cars to be equipped with the eCall technology from April 2018 [30]. In the event of a road accident, an eCall equipped vehicle will automatically (or manually) establish an emergency voice call that is routed to the appropriate Public Safety Answering Point (PSAP). The eCall equipped vehicle also sends an emergency message known as the minimum set of data (MSD), including key information about the accident, such as number of vehicle occupants, time, accurate location, driving direction resulting from accurate satellite-based data and vehicle description. The current eCall is based on Circuit Switched (CS) emergency calls in GSM and UMTS networks and therefore is also dependent on the assurance of those networks with uninterruptible power supplies (UPS) in case of a power outage in a disaster situation and may work only for a few hours. [31]

For the immediate needs of communication after the occurrence of a disaster, approaches that use the infrastructure which is still available ([3], [5], [8]) seem to be most suitable. It should be emphasized that non of these solutions is focusing on the lifetime of the network which, however, is a key part of this thesis.

The following paragraphs will focus on suitable applications that can provide the required services in disaster areas. The importance of immediate availability of a communication network after occurrence of a disaster can be strengthened by the type of service application that victims and rescuers immediately want to use. An example of an application for messaging even without Internet connection is FireChat [32] which is a free messaging App available for Android and iOS. FireChat can be used for public and private communications that works even without Internet access or cellular data. When there is no Internet connection or cellular networks available, FireChat uses bluetooth or Wi-Fi to connect directly from phone to phone by a custom proprietary ad-hoc protocol. If there are more than two devices FireChat is forming a network. In order to overcome the limitations in the ad-hoc protocol, forwarding is done at the application layer. The App has recently been used in pro-democracy protests in Taiwan (April 2014), Hong Kong (September 2014) and natural disasters including floods in Kashmir (April 2015) and Chennai (October 2015). During the protests in Hong Kong's "umbrella movement" the App has been downloaded 100.000 times in 24 hours. [32]
FireChat Alerts allows governments to communicate with populations during emergencies such as natural disasters or health crises. The messaging App has been evaluated and further compared to traditional communication networks in a case study in the framework of the earthquake preparedness initiative carried out by the Metro Manila Development Authority at the Philip-

pines during June 2016. The Metro Manila Shake Drill (MMShakeDrill) was one of the world's largest earthquake preparedness initiatives. The MMShakeDrill simulated a huge earthquake whereby traditional communication networks are rapidly congested and physical infrastructure will be severely damaged, making the dissemination of critical information extremely challenging and complicating rescue operations. During and after the MMShakeDrill, the authority sent alerts to the population across Metro Manila using the broadcast messaging capabilities of MeshKit in FireChat Alerts. The study concludes that at a density of over 700 users/km^2, only 32% of users received a message from servers and cell towers, whereby the FireChat MeshKit delivered messages to 80% more users [32].

It can be concluded that FireChat would be an ideal application to be used by people in need in disaster situations.

The Federal Office of Civil Protection and Disaster Assistance – Bundesamt für Bevölkerungsschutz und Katastrophenhilfe (BBK) – in Germany has also developed an Emergency Information and Warning App called NINA [13]. This App is intended to warn users about emergencies and hazards all over Germany, such as severe weather, floods and other relevant events. The push functionality calls the attention of the users always to current threats. However, it has to be emphasized that during an electric power breakdown, mobile phones will only continue to work for a short time and the mobile communication and internet access is no more available. A broadcast messaging capability such as used by MeshKit in FireChat Alerts is currently not available.

The smart phone App "smarter" [14] represents a promising approach which is still under development and so far only available for android smart phones. A case study carried out in September 2017 revealed that nearly 70 percent of the participants would download the smarter App in case of a disaster on their smart phones. Therefore the smarter architecture has been implemented in the Serval Mesh^5 open-source disaster communications system [33]. The smarter architecture provides also the data services built on top of the communication layer, e.g., SOS messages in disaster situations [34]. However, the lifetime which relies on the batteries of the smart phones is still an unsolved issue. Recently also other chat App's have been developed which made use of the Multipeer Connectivity Framework that appeared in apple iOS 7.

Both mentioned App's don't rely on the IEEE 802.11 mesh network amendment which is used in this work. FireChat is based on ad-hoc and Serval Mesh is based on the AP and Client mode. Thus both need an additional forwarding protocol on higher layers which is further discussed in Section 2.2.3.3. The lifetime of the network and of the user nodes is not investigated in both solutions. It should be noted that the lifetime enhancement algorithm which is proposed in this work (see Chapter 5) can be realized with particular ease by using the IEEE 802.11s Mesh Peering Management (MPM) protocol and without additional protocol effort.

---

^{5}The Serval Mesh Project is an open-source mesh communications platform that arose in response to the Haiti Earthquake of 2010. The Serval Mesh software allows off-the-shelf smart phones to build self-organized mesh networks using Wi-Fi in AP and client modes and a store-and-forward protocol (Serval Rhizome). [33]

2.2 IEEE 802.11 Wireless Local Area Networks

This section describes only those parts of the IEEE 802.11 standard which are relevant for this work mainly focusing on the amendment for mesh networking IEEE 802.11s introduced in 2011[6]. The current version is IEEE 802.11-2016 [35] where further details on the topics briefly summarized here can be found. The IEEE 802.11 standard specifies the Physical, Medium Access Control (MAC) and Link Layer operations for wireless LANs. It can be distinguished between three network architectures: the infrastructure Basic Service Set (BSS), the Independent BSS (IBSS) which is also referred to as ad-hoc network and the Mesh Basic Service Set (MBSS).

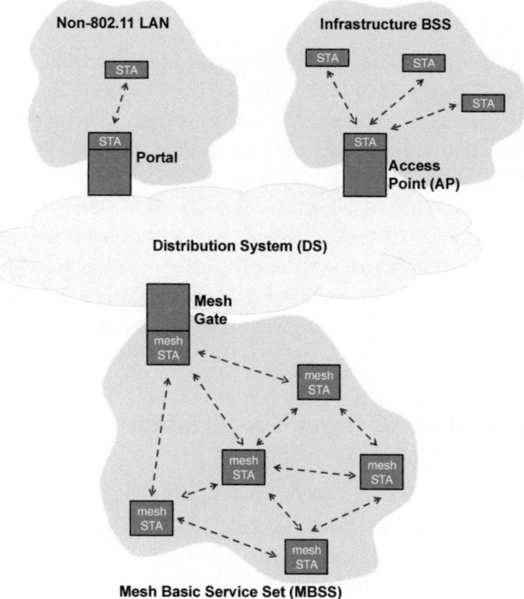

Figure 2.1: 802.11 architectures inspired by [35, Figure 4-9] containing Stations (STAs), mesh STAs (MSTA), Mesh Gate, Access Point (AP) and Portal.

Figure 2.1 shows an example for mesh and infrastructure BSSs. Only mesh STAs participate in a mesh BSS and use the associated functionalities such as formation of the mesh BSS, path selection and forwarding. A mesh STA is not a member of an IBSS or an infrastructure

---

[6]IEEE specific terms and message types use exactly the spelling as used in the standard [35], including upper and lower case spelling.

BSS and does not communicate with non-mesh STAs. However, a MBSS may also interconnect with other BSSs or MBSS through the so called Distribution System (DS). The DS is a system used to interconnect a set of basic service sets (BSSs) and integrated LANs to create an Extended Service Set (ESS). In that way mesh STAs may also communicate with non-mesh STAs via a Mesh Gate. Access to distribution services via the wireless medium for associated STAs in an infrastructure BSS is given by an Access Point (AP). Additionally, a portal integrates the IEEE 802.11 architecture with a non-IEEE-802.11 LAN (e. g. a traditional wired LAN). Additional information about the 802.11 architectures are given in [35, Chapter 4].

2.2.1 IEEE 802.11 Physical Layer Specification

The IEEE 802.11 standard defines different Physical Layers (PHYs) which are provided to the Medium Access Control (MAC) layer, notably Direct Sequence Spread Spectrum (DSSS), High Rate DSSS (HR/DSSS), Orthogonal Frequency-Division Multiplexing (OFDM) or High Throughput OFDM (HT/OFDM). This work is based only on the OFDM PHY specification described in [35, Chapter 17]. The OFDM system provides a WLAN with data payload communication capabilities of 6, 9, 12, 18, 24, 36, 48, and 54 Mbit/s. The system uses 52 subcarriers out of a 64 subcarrier IFFT/FFT system that are modulated using Binary or Quadrature Phase Shift Keying (BPSK or QPSK) or using 16- or 64-Quadrature Amplitude Modulation (16-QAM or 64-QAM). Forward error correction coding (convolutional coding) is used with coding rates of 1/2, 2/3, or 3/4. Additional information is given in [35, Chapter 17].

2.2.2 IEEE 802.11 MAC Layer Specification

Due to the distributed nature of the MBSS, only the Mesh Coordination Function (MCF) is present in a mesh STA. A mesh STA thereby exclusively uses the MCF for channel access. The MCF provides two methods for channel access: The mandatory Enhanced Distributed Channel Access (EDCA) and the optional MCF Controlled Channel Access (MCCA). In an MBSS each mesh STA is a Quality of Service (QoS) STA with QoS extensions originally described in IEEE 802.11e[7]. The EDCA mechanism allows the contention-based access to the medium and is using the Carrier Sense Multiple Access/Collision Avoidance (CSMA/CA) scheme. EDCA differentiates between four access categories (ACs) for different user priorities (UPs). The four ACs are AC_BK for background traffic, AC_BE for Best Effort traffic, AC_VI for Video traffic and AC_VO for Voice traffic (see Appendix A for further information). Figure 2.2 shows the four ACs representing four priorities which are individually using one of the four independent backoff entities. These backoff entities are essentially controlled by the contention window

---

[7]802.11e amendment for Medium Access Control (MAC) Quality of Service Enhancements.

(CW) parameter and the so called inter frame spacing. The contention window parameter is an integer within the range of values of the PHY characteristics aCWmin and aCWmax, aCWmin $\leq$ CW $\leq$ aCWmax. According to the AC further restrictions to the CW parameter apply (see Appendix A).

The IEEE 802.11 MAC is using different interframe gaps, denoted as Inter Frame Spaces (IFS) in order to control the medium access. Some of the most common IFS are short interframe space (SIFS), PCF[8] interframe space (PIFS), DCF[9] interframe space (DIFS) and arbitration interframe space (AIFS) (which is used by the QoS facility). The arbitration interframe space denotes the time interval between QoS data frames. Thereby the contention window (CW) parameter and the arbitration interframe space depend both on the used AC and provide for the basic mechanism of QoS provisioning. ACs with lower priority are attributed longer backoff times by the help of those two parameters. Additional information is given in [35, Chapter 10].

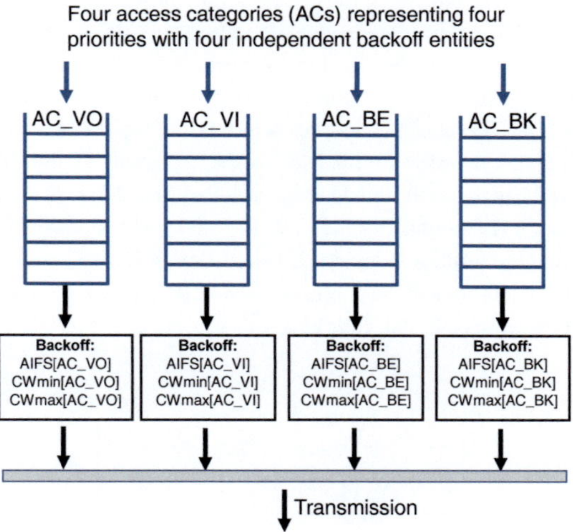

Figure 2.2: 802.11e EDCA according to [35, Fig. 10-24].

2.2.3 IEEE 802.11s Amendment for Mesh Networking

The IEEE 802.11s amendment for mesh networks describes only changes to the MAC-layer, but not to the PHY. It is also compatible with higher layer protocols, e.g. the mesh security

---

[8]Point Coordination Function [35, Chapter 10].

[9]Distributed Coordination Function [35, Chapter 10].

is based on IEEE 802.11i. Most important are the Hybrid Wireless Mesh Protocol (HWMP) which is defined as the default path selection protocol for the mesh BSS and the Mesh Peering Management (MPM) protocol which is used to establish, maintain and close mesh peerings between mesh STAs. Compared to classic routing protocols the routing in IEEE 802.11s is performed on layer 2 and therefore the routing protocol uses MAC addresses instead of IP addresses. The preferred term for routing in IEEE 802.11s is path selection, in order to make the difference to routing on layer 3 with IP addresses more distinct.

2.2.3.1 Peer link establishment

Figure 2.3 shows the minimum messaging required for peer link establishment according to the MPM. In order to join a mesh network a mesh STA has to establish peer links to all its neighbor mesh STAs (at least one of them). This is carried out by two two-way-handshake mechanisms (one in each direction of the link) with the corresponding neighbor MSTA to which a peer link has to be established. Beacon frames are transmitted periodically by each MSTA to announce the presence of the MSTA to others. Having received a beacon frame, a MSTA then sends a Mesh Peering Open message which is replied by a Mesh Peering Confirm message. The involved MSTA also sends a Mesh Peering Open message which is also replied by a Mesh Peering Confirm message and the peer link between the two MSTAs is established. Figure 2.3 shows an example of the peer link establishment. It should be noted that the order of the messages does not need always to be as in the example of Fig. 2.3. Especially, the number of messages might be higher due to interference leading to lost packets and thus requiring retransmission of packets. The whole peer link establishment, maintenance and close process is shown in the finite state machine of the MPM protocol in [35, Fig. 14-2].

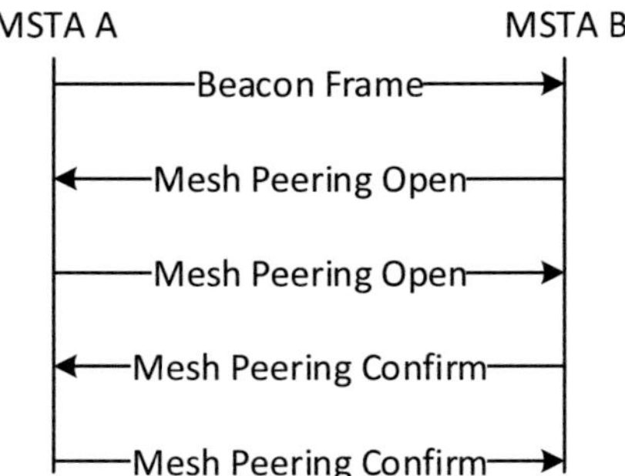

Figure 2.3: Two-way-handshake Peer Link establishment according to [35, Fig. 14-2].

2.2.3.2 Path selection protocol

HWMP provides both proactive path selection (proactive tree building mode) and reactive path selection (on-demand mode). On-demand path selection offers a higher flexibility in changing environments whereby the pro-active tree based path selection is very efficient in fixed hierarchical mesh deployments with dedicated root nodes. The path selection protocol uses link metrics in the assessment of a mesh path to the destination. The airtime link metric described in [35, Chapter 14.9] is the default link metric. The airtime c_a for each link can be calculated as follows:

$$c_a = \left[O + \frac{N_B}{R} \right] \frac{1}{1 - e_f} \qquad (2\text{-}1)$$

O varies depending on the PHY and is dependent among others on the channel access overhead which includes frame headers, training sequences or access protocol frames. N_B denotes the number of bits in the test frame with a recommended length of 8192 bits according to [35, Tab. 14-4]. R represents the data rate (in Mb/s) at which the mesh STA would transmit a frame of standard size N_B based on current conditions. e_f is the frame error rate for the test frame of size N_B.

The path selection protocol which is activated in the MBSS is indicated by the Active Path Selection Protocol Identifier field. The Hybrid Wireless Mesh Protocol is set as default path selection protocol in either proactive or reactive path selection mode. The mesh routing architecture is extensible whereby the active path selection protocol can also be specified in a vendor specific element [35, Section 9.4.2.98]. The foundation of HWMP is an adaptation of the reactive Ad-hoc On-demand Distance Vector routing protocol (AODV) [36, RFC 3561]. In contrast to AODV, HWMP works on layer 2 with MAC addresses and uses a radio-aware routing metric for the path selection while AODV works on layer 3 with IP addresses and uses the hop count as routing metric [37]. Additional information is given in [35, Chapter 14.2].

The proactive extension of HWMP uses the same distance vector methodology and reuses some of the routing control messages. If a root portal is present, a distance vector routing tree is built and maintained. Thereby the tree based routing avoids an unnecessary flooding with messages during discovery and recovery. The presence of a MSTA configured as root MSTA is announced by sending periodically Root Announcement (RANN) messages. It is clear, that the scalability in such a centralized approach is limited.

In this work, it is assumed that no root portal is present and that the mesh network is not hierarchical. Therefore only the on-demand mode of HWMP will be used. Important messages are PREQ (path request), PREP (path reply) and PERR (path error). The PREQ element is used for discovering a path to one or more target mesh STAs, maintaining a path (optional in IEEE 802.11) and confirming a path to a target mesh STA (optional in IEEE 802.11)[10].

---

[10]The PREQ element is used also for building a proactive (reverse) path selection tree to the root mesh STA which is out of scope of this work using only the on-demand mode of HWMP.

The PREQ element is transmitted in an HWMP Mesh Path Selection frame [35]. The PREP element is used to establish a forward path to a target and to confirm that a target is reachable. The PREP element is issued in response to a PREQ element. The PERR element is used for announcing an unreachable destination. See [35, Sec. 14.10] for further information.

2.2.3.3 Alternative mesh routing protocols

The Radio-Aware Optimized Link State Routing protocol (RA-OLSR) was originally foreseen as an optional proactive routing protocol for the IEEE 802.11s framework, however, it was not included in the final amendment. Therefore it can be seen as an alternative routing protocol to HWMP which can be implemented by vendors. RA-OLSR also uses MAC addresses and a radio-aware metric instead of IP addresses and the hop count metric originally present in the Optimized Link State Routing Protocol (OLSR) [38, RFC 3626].

Additional routing protocols to be mentioned are BABEL [39, RFC 6126] or B.A.T.M.A.N. (Better Approach To Mobile Ad-hoc Networking [40]), however, a lot more are available.

The BABEL routing protocol [39, RFC 6126] is also a distance-vector routing protocol for IP packet-switched networks which is based on the ideas proposed in Destination-Sequenced Distance Vector routing (DSDV), AODV, and Cisco's Enhanced Interior Gateway Routing Protocol (EIGRP), but it uses different techniques for loop avoidance [41]. One limitation is that BABEL relies on periodic routing table updates rather than using a reliable transport protocol. Hence, in large, stable networks it generates more traffic than protocols that only send updates when the network topology changes.

B.A.T.M.A.N. is a layer 3 proactive routing protocol for multihop ad-hoc mesh networks. B.A.T.M.A.N.-advanced (often referenced as batman-adv) is an implementation of the original B.A.T.M.A.N. routing protocol in form of a linux kernel module operating on layer 2 [40]. Originally B.A.T.M.A.N. was developed as a replacement or alternative to OLSR and became famous also in the freifunk community [42] which is using both, B.A.T.M.A.N. and OLSR. B.A.T.M.A.N. was originally based on the IBSS ad-hoc mode, but recently it is also used in conjunction with IEEE 802.11s.

The efficiency of different network routing and path selection protocols for wireless mesh networks of the five most popular protocols, namely Ad-Hoc On Demand Distance Vector (AODV), Dynamic Manet On Demand (DYMO), HWMP, OLSR and B.A.T.M.A.N. has been studied in [43] by network simulations with OMNeT++ [44]. The pro-active protocols OLSR and B.A.T.M.A.N. showed a lower packet loss than the reactive protocols AODV, DYMO and HWMP. However, DYMO and HWMP are characterized by a packet delivery delay of more than three times smaller than shown by the AODV protocol and comparable to the B.A.T.M.A.N. routing protocol. The energy consumption, with respect to AODV, is reduced by 30% for HWMP and 20% for OLSR. The usage of the proactive routing protocols require continuous exchange of information between network nodes, which translates into a constant energy consumption of OLSR and B.A.T.M.A.N. about 20% higher than HWMP.

2.2.3.4 Conclusion

In this work only HWMP of IEEE 802.11 will be used and analyzed. The main reason is the availability in devices as HWMP is the only protocol mandatory according to IEEE 802.11. Interoperability in terms of integrating different MSTAs into a single WMN would also be not possible when using different routing or path selection protocols.
The performance of HWMP and batman-adv has been analyzed e. g. in [45] in terms of route stability and recovery times. The open802.11s [46] implementation which has been used showed some instabilities, but recovered quite rapidly in case of node failures. On the other hand batman-adv had problems in resuming the communication after an abrupt interruption according to [45]. However, the implementation of 802.11s [46] which has been tested was based on a draft version and also batman-adv is under ongoing development. It was shown that both protocols have strengths and shortcomings.

The choice of an appropriate routing or path selection protocol is also dependent on the use case and topology. The HWMP on-demand mode works better in dynamic networks and rapidly changing network topologies. The reactive path selection protocol has a lower overhead compared to the proactive protocols, because the path is only discovered when needed and paths which are never used are not maintained. Therefore HWMP reduces traffic overheads. This holds also for the DRS proposed in this work which needs to enable only a low number of contemporaneous emergency calls as well as text messaging whereby a path is only discovered when needed. Additionally, the proposed Lifetime Enhancement Algorithm (LEA) described in Chapter 5 results in a changing network topology because of the deactivation of redundant MSTAs and reactivation of MSTAs when needed which makes proactive routing protocols inappropriate.
Finally, the advantage of path selection protocols (layer 2) compared to routing protocols (layer 3) such as BABEL or OLSR is that the MAC layer has timely accurate information on network topology, link-quality, data rates and queue lengths.

Chapter 3

Evaluation Methods for Coverage and Network Performance of Wireless Mesh Networks

Contents

The following chapter provides evaluation methods for the coverage and network performance of Wireless Mesh Networks (WMNs). First, state-of-the-art methods for the calculation of the coverage will be presented with particular focus on their usage in disaster areas. Second, the evaluation methods for the performance estimation of WMNs which are based on network simulations will be elaborated. Finally, the throughput and delay performance will be shown based on state-of-the-art research of WMNs.

3.1 Coverage, Connectivity and Lifetime of Wireless Mesh Networks

Among the key considerations concerning the performance of a DRS are the number of nodes needed to set up a network in the specified area[1], the coverage that could be obtained by the DRS within the specified area by a given number of nodes, the scalability of the DRS to larger or smaller areas as well as the lifetime of the DRS.

Especially in the field of wireless sensor networks there exists a considerable number of research results on those topics, which could also be considered for wireless mesh DRS and which will be discussed in the sections concerning the lifetime of the DRS.

3.1.1 Coverage

The coverage of the whole DRS which is realized as Wireless Mesh Network (WMN) is dependent on the coverage of each station which can be estimated in a link budget calculation. We thus start with the analysis of appropriate methods for link budget calculations for DRSs based on WMNs. To do so path loss as well as coverage probability has to be considered. In order to correctly account for the path loss a suitable model depending on the propagation scenario has to be chosen. Random variations of the channel due to multipath effects or shadowing are taken into account by adding appropriate margins.

3.1.1.1 Path loss

Path loss is the ratio (generally expressed in dB) between the transmitted power $\mathcal{P}_{\mathrm{Tx}}$ and the received power $\mathcal{P}_{\mathrm{Rx}}$ of a communication link. In general it can be distinguished between stochastic and deterministic path loss models. Usually, stochastic path loss models are found on basis on empirical or additional theoretical assumptions. Empirical models are based on extensive measurement data. These models are simple, however, not very accurate. Semi-empirical models are based on empirical models with deterministic aspects (e.g. heights of buildings, widths of roads). Purely deterministic models are site-specific and require an enormous amount of geometry information about the site which may lead to very accurate models (e.g. based on ray tracing algorithms). However, the computational effort for purely deterministic models is very high which makes them inappropriate for this work. Empirical and semi-empirical models have been studied in several research papers and all of them are designed for different environments (e.g. outdoor, indoor, urban or rural) and different frequencies. Some of the most

---

[1]More precisely: what actually has to be studied is the required node density.

common semi-empirical path loss models are Okumura [47], Okumura-Hata [48], Cost[2]-231 Hata [49], ECC-33 [50], Erceg [51] and the Stanford University Interim (SUI) Model [52]. All these models are restricted to frequencies below 5 GHz. This work, however, focuses on the 5 GHz band used by IEEE 802.11 a/n. Higher frequencies have been studied e. g. in [53] where an evaluation of a deployed WiMAX system operating in the 4.9 GHz Public Safety Band in the City of Tulsa, USA is presented. Existing propagation models in the UHF band (300 MHz to 3 GHz) were applied to 4.9 GHz and the characteristics predicted by the models were compared with the measured data. The three models which have been compared were the SUI Model, the Cost-231 Hata Model and the ECC-33 Model. The results showed that out of the three models presented, the Cost-231 Hata Model was able to predict best the characteristics of the deployed WiMAX system.

Another family of semi-empirical path loss models are the ITU[3] IMT[4]-advanced channel models [54]. The ITU IMT-advanced channel models are based on extensive measurements and cover a large range of propagation scenarios and environments, including indoor-to-outdoor, outdoor-to-indoor, urban micro-cells, urban macro-cells and rural macro-cells. The models can be applied in the frequency range from 2 to 6 GHz and for different antenna heights. The IMT-advanced channel models will be used and further analyzed in this work rather than the Cost-231 Hata model which was originally not considered for frequencies above 4 GHz. Other reasons compared to the previously mentioned models are the validity of the IMT-advanced models up to 100 MHz RF bandwidth where the bandwidth of the COST-231 Hata model was aimed for 2G/3G cellular systems with up to 5 MHz bandwidth as well as the larger range of propagation scenarios. As the considered DRS is based on propagation ranges of the order of magnitude of 100 m and a bandwidth of 20 MHz, both stemming from the key technical parameters of IEEE 802.11a/n OFDM systems, it is clear that the IMT-advanced channel models are more appropriate.

Path loss models which cover the propagation through rubble (e. g. after an earthquake) are in general not available. It is assumed that the propagation in a catastrophic area is mainly of Non-Line-of-Sight (NLOS) type which means that only a certain number of weak multipath echoes are received and no direct signal is available. Wi-Fi equipped cars parking along the roadside or battery powered MSTAs which have not been destroyed during the disaster will have connection to others in this Non-Line-of-Sight area. Further it is assumed that the 5 GHz band is mainly used as there are more frequency channels available and the band might be less crowded by other systems compared to the 2.4 GHz Industrial, Scientific and Medical Band (ISM) even in a hard to predict disaster recovery situation. The urban micro-cellular environment considered as a propagation environment within the IMT-advanced channel model family and which assumes a high user density is focusing on pedestrian and slow vehicular users and

---

[2]COST is an intergovernmental framework for European Cooperation in Science and Technology, allowing the coordination of nationally-funded research on a European level.

[3]International Telecommunication Union (ITU).

[4]International Mobile Telecommunications (IMT).

therefore suits best to the disaster recovery scenario of this work. Additionally, the height of both the antenna at the base station (BS) and that at the user terminal (UT) is assumed to be well below the tops of surrounding buildings for this model [54]. The IMT-advanced urban micro cell NLOS path loss model [54] is therefore assumed to be appropriate in a disaster area in an urban environment.

Table 3.1 shows the path loss (PL) for the Urban Micro cell (UMi) model [54] for distance d (in m) and center frequency f_c (in GHz). Compared to the UMi path loss model a second model for rural environments will be evaluated. The probability for Line-of-Sight (LOS) conditions in the rural area is higher than in an urban area. The Rural Macro (RMa) cell Line-of-Sight (LOS) model therefore might be better to cover disaster areas in rural environments.

Table 3.1: Comparison of the IMT-advanced channel models considered in this work. [54, Tab. A1-2]

Model	Path loss (PL) in dB (f_c in GHz, distance d in m)	Shadow fading std[5] in dB	Applicability range, antenna height default values
UMi NLOS	Hexagonal cell layout: $\mathrm{PL} = 36.7 \log_{10}(d) + 22.7 + 26 \log_{10}(f_c)$	$\sigma_F = 4$	$10\,\mathrm{m} < d < 2000\,\mathrm{m}$ $h_{\mathrm{BS}} = 10\,\mathrm{m}$ $h_{\mathrm{UT}} = 1 - 2.5\,\mathrm{m}$
RMa LOS	$\mathrm{PL}_1 = 20 \log_{10}(40\pi\, d\, f_c/3)$ $+\min(0.03h^{1.72}, 10) \log_{10}(d)$ $-\min(0.044h^{1.72}, 14.77) + 0.002 \log_{10}(h)d$ $\mathrm{PL}_2 = \mathrm{PL}_1(d_{\mathrm{BP}}) + 40 \log_{10}(d/d_{\mathrm{BP}})$	$\sigma_F = 4$ $\sigma_F = 6$	$10\,\mathrm{m} < d < d_{\mathrm{BP}}$ [6] $h = $ avg. building height $W = $ street width $d_{\mathrm{BP}} < d < 10000\,\mathrm{m}$ $h_{\mathrm{BS}} = 35\,\mathrm{m}, h_{\mathrm{UT}} = 1.5\,\mathrm{m},$ $W = 20\,\mathrm{m}, h = 5\,\mathrm{m}$ Applicability ranges: $5\,\mathrm{m} < h < 50\,\mathrm{m}$ $5\,\mathrm{m} < W < 50\,\mathrm{m}$ $10\,\mathrm{m} < h_{\mathrm{BS}} < 150\,\mathrm{m}$ $1\,\mathrm{m} < h_{\mathrm{UT}} < 10\,\mathrm{m}$

Table 3.1 shows the applicability ranges of the Urban Micro cell (UMi) NLOS model and Rural Macro (RMa) LOS model, e. g. average building height h and street width W according to [54, Tab. A1-2]. In the urban micro-cell scenario the height of both the antenna at the base station (BS) and at the user terminal (UT) are assumed to be well below the tops of surrounding

---

[5] std denotes the standard deviation of the log normal distribution that is used to model the attenuation due to shadowing effects.

[6] The break point distance is given by $d_{\mathrm{BP}} = 2\pi \cdot h_{\mathrm{BS}} \cdot h_{\mathrm{UT}} \cdot f_c/c$, where f_c is the centre frequency in Hz, $c = 3.0 \cdot 10^8\,\mathrm{m/s}$ is the propagation velocity in free space, and h_{BS} and h_{UT} are the antenna heights at the base station (BS) and the user terminal (UT), respectively.

buildings which is also the case for a wireless mesh disaster recovery topology. In contrast to a mobile radio cell there is no difference between BS and UT in a wireless mesh network, however, the nomenclature of the IMT-advanced models is retained. The default height of the base station $h_{BS} = 10\,\text{m}$ would be too high in the mesh network scenario proposed here. The IMT-advanced channel models on the other hand are based on the WINNER[7] II channel models [55] whereby the modeling methodology and channel construction between WINNER II and ITU are the same. In [55] a height of the base station h_{BS} of the following range is proposed for the UMi model: $5\,\text{m} < h_{BS} < 20\,\text{m}$. This height suits for MSTAs located at buildings, but not for users and MSTAs in cars. For the DRS proposed in this work, so called STA to STA links with a height of the antenna $h_{BS} = h_{UT} = 1.5\,\text{m}$ and AP to AP links with a height of the antenna $5\,\text{m} \leq h_{BS} = h_{UT} \leq 20\,\text{m}$ as well as AP to STA links exist beside each other.

Currently the IEEE is developing an amendment for High Efficiency Wireless Local Area Networks (HEW)[8] (sometimes called Next Generation Wi-Fi) denoted as IEEE 802.11ax. The main focus is on increasing capacity and improving spectral efficiency, e. g. by the use of multi-user operation by means of Multi User Multiple-Input/Multiple-Output (MU-MI/MO) and Orthogonal Frequency-Division Multiple Access (OFDMA) technologies specified for both downlink and uplink, larger FFT size (2048), smaller subcarrier spacing (78.125 kHz) and higher modulation scheme (1024-QAM). Additionally, Device to Device (D2D) communications are foreseen as possible extension. The use of D2D communications will reduce the amount of airtime required for each transmission by allowing higher data rates and avoiding the use of the AP as a relay. Adjacent STAs in close range can transmit data via direct links. The ITU IMT-advanced Urban Micro channel model was agreed by the HEW study group as a consensus model for system level simulations of AP to STA, AP to AP and STA to STA links [57]. The height parameters thereby have been modified depending on the link, i. e. for AP to AP links $h_{BS} = h_{UT} = 10\,\text{m}$ and for STA to STA links $h_{BS} = h_{UT} = 1.5\,\text{m}$. While the propagation parameters for the UMi model remain the same it is recommended to change the shadow fading standard deviation σ_F from $4\,\text{dB}$ to $7\,\text{dB}$ for STA to STA links in order to account for the higher impact of shadowing near the ground level. The shadow fading standard deviation σ_F of log-normal shadowing for DRS applications will be further considered in the following section.

Following these findings, the ITU Urban Micro cell (UMi) model will be used throughout this work in order to predict the path loss for system level simulations of the disaster recovery wireless mesh network in urban areas.

The RMa LOS model which is also shown in Tab. 3.1 will not be further used throughout this work, as we will focus on urban areas only. Another reason is the applicability range of the model which is aimed on mobile radio cellular scenarios with a higher default height of the BS compared to the UMi model which does not cover STA to STA links and thus makes the RMa LOS model questionable for the evaluation even of DRS in rural environments.

---

[7]WINNER II, Project ID: 027756, Funded under FP6-IST, IST-2004-2.4.5 – Mobile and Wireless Systems and Platforms Beyond 3G.

[8]The High Efficiency WLAN Study Group (HEW SG), a study group within the IEEE 802.11 working group, is considering improvement of spectrum efficiency to enhance the system throughput/area in high density scenarios of APs and/or STAs. [56]

3.1.1.2 Coverage probability

The coverage probability, more precisely the outage probability P_{out} shown in Eq. (3-1), describes the probability that a wireless link is not feasible even if the distance between transmit and receive antenna is in a range where the mere path loss model predicts availability of the link. $\mathcal{P}_{\text{Rx,Th}}$ is the receiver threshold.

$$P_{\text{out}} = P_r \left(\mathcal{P}_{\text{Tx}} - \text{Pathloss} > \mathcal{P}_{\text{Rx,Th}} \quad \wedge \ \text{link not feasible} \right) \tag{3-1}$$

$$P_{\text{coverage}} = 1 - P_{\text{out}} \tag{3-2}$$

This is happening due to shadowing and multipath propagation which both are responsible for so called fading effects (Fig. 3.1). A fading margin thus has therefore to be included in the link budget which accounts for the fading relevant at a given coverage probability. We thereby have to distinguish between shadow fading and multipath fading. For both types of fading different fading margins have to be included into the link budget. Both margins will be discussed for the DRS under consideration through this section.

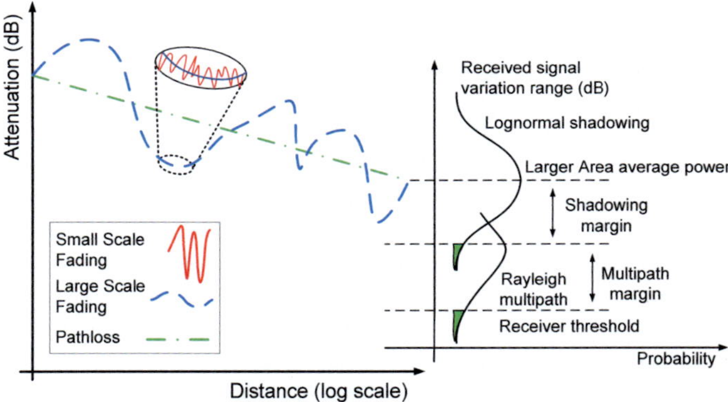

Figure 3.1: Sketch of received signal variation range (in dB) due to path loss, large-scale and small-scale fading.

Shadow fading also referred to as large-scale fading is a phenomenon that may occur when the transmitter or receiver is behind an obstruction and the radio link experiences a significant reduction in signal power, but still allowing for connection. If the surface of the objects which causes the obstruction is smooth, part of the signal is reflected and a part of the signal penetrates the objects and thus can be transmitted through the object. If the surfaces are rough the waves are diffusely scattered. Additionally, waves can be diffracted at the edges of objects. Multipath fading refers to the dramatic changes in signal amplitude and phase that can be experienced as a result of small changes (as small as half a wavelength) in the spatial position between transmitter and receiver due to constructive or destructive interference of the multipath components contributing to the link [58].

Shadow fading is modeled as log-normal distribution and therefore the probability density function (PDF) of the logarithmic fieldstrength values is a standard Gaussian distribution. The corresponding cumulative density function (CDF) is thus given by a Q-function defined as: [58]

$$Q(a) = \frac{1}{\sqrt{2\pi}} \int\limits_{a}^{\infty} e^{\left(-\frac{x^2}{2}\right)} \, \mathrm{dx} \tag{3-3}$$

The outage probability due to shadow fading is calculated as follows [58]:

$$P_{\mathrm{out}} = Q\left(\frac{M_{\mathrm{large\_scale_{dB}}}}{\sigma_F}\right) \tag{3-4}$$

$M_{\mathrm{large\_scale_{dB}}}$ denotes the margin which accounts for the shadow fading at a given coverage probability. Inserting the standard deviation $\sigma_F = 4\,\mathrm{dB}$, e.g. as given by the path loss model UMi NLOS (see Tab. 3.1) or $7\,\mathrm{dB}$ as recommended for STA to STA links and $P_{\mathrm{out}} = 0.1$ – assuming a 90% coverage – into Eq. (3-4) we get:

$$\begin{aligned}
M_{\mathrm{large_scale_{4dB}}} &= 4\,\mathrm{dB} \cdot Q^{-1}(0.1) = 4\,\mathrm{dB} \cdot 1.282 = 5.13\,\mathrm{dB} \\
M_{\mathrm{large_scale_{7dB}}} &= 7\,\mathrm{dB} \cdot Q^{-1}(0.1) = 7\,\mathrm{dB} \cdot 1.282 = 8.97\,\mathrm{dB}
\end{aligned} \tag{3-5}$$

Small-scale fading may occur either due to time delay spread resulting from different propagation delays for the macroscopic multipath components of the channel occurring from scatterers at large distances from sender and receiver or due to Doppler spread resulting from different Doppler shifts for the microscopic multipath components which are received from scatterers in the vicinity of sender or receiver. The wireless channel thus can be both frequency-selective[9] as well as time-selective[10]. The calculation of the coherence bandwidth B_c and the coherence time T_c is shown in [59, pp. 163–166]. It should be noted that the rms (root mean square) delay spread and coherence bandwidth are inversely proportional, i.e. the higher the delay spread, the lower the coherence bandwidth and therefore the higher the frequency selectivity of the channel. Doppler spread and coherence time are inversely proportional as well.

As already mentioned the cause of frequency-selective channels is delay spread. This can be easily understood when noting that a delay spread in the order of the symbol duration T_{sym} of the considered communication system causes significant variation of the received field strength over the system bandwidth (being proportional to $1/T_{\mathrm{sym}}$). The negative impact of frequency-

---

[9]Usually a channel is said to be frequency-selective if the coherence bandwidth of the channel B_c is smaller than the signal bandwidth B_{sig}.

[10]Time selective fading can occur on rather different time scales. For most channels and communication systems coherence time is considerably larger than the frame duration which means that channel estimation usually can be accomplished by standard procedures. This also means that often the average duration of deep fades is long enough such that complete frames may be lost. For mobile communication systems this can be compensated by methods such as Hybrid Automatic Repeat Request (HARQ) or Rx diversity concepts.

selective fading on one hand thus can be attributed to inter-symbol interference (ISI) created by the differences in propagation delay of the macroscopic multipath components. On the other hand a frequency-selective channel may lead to significant small-scale fading effects, even if ISI is negligible because there is specific fading on the individual subcarriers. Let us first consider the effect of ISI for the case of the proposed DRS. The communication channel can only be assumed to create neglectable ISI if the symbol duration T_{sym} is much longer than the rms delay spread $\Delta\tau$ ($T_{\mathrm{sym}} \gg \Delta\tau$). In an OFDM system inter-symbol-interference even can be completely avoided when using an appropriate guard interval during which the signal of the OFDM symbol is cyclically extended. The symbol duration T_{sym} of the IEEE 802.11a OFDM PHY according to [35, Tab. 17-5] is $4\,\mu$s where the guard interval duration T_{GI} is $0.8\,\mu$s. This also shows the advantage of an OFDM multi-carrier system with increased symbol duration for each of the subcarriers that are transmitted in parallel when compared to a single carrier system where the single carrier occupies the whole bandwidth leading to a much shorter symbol duration. The maximum delay spread for which inter symbol interference (ISI) is prevented in the IEEE 802.11a OFDM PHY therefore is $0.8\,\mu$s. The time difference of $0.8\,\mu$s corresponds to a path length difference of $240\,$m. It should be noted that the delay spread in the outdoor environment in a disaster area might be even higher due to destroyed buildings and a larger number of reflecting surfaces, which will further be discussed below.

Multipath channels with time delay spread can be modeled by so called tapped delay line or cluster delay line models simplifying the impulse responses of the real channel for a finite number of impulses N_d (i. e. a model which represents the channel by a delay line with N_d taps). At this point it is worth noting the advantage of the IMT-advanced models, namely that they do not only provide mere path loss models, but provide the parameters for modelling the corresponding mobile channels as time variant linear systems with the help of cluster delay line models as well. The clustered delay line model of the IMT-advanced UMi NLOS model is shown in [54, Tab. A1-12]. For the taps with higher delay than $440\,$ns shown in [54, Tab. A1-12] a power level of the taps less than $-10\,$dB is shown. By using the equations in [59, p. 160] a rms delay spread for the UMi NLOS model of $\Delta\tau = 130\,$ns can be calculated which is less than the guard interval duration $T_{\mathrm{GI}} = 0.8\,\mu$s. Nevertheless, there will be substantial fading on the individual subcarriers as mentioned above.

Additionally, [60] shows results of an extensive field measurement campaign involving vehicular wireless channels across a wide variety of speeds and propagation conditions (LOS and NLOS) which focusses on dedicated short-range communications based on IEEE 802.11p[11]. These results can also be taken into account for the disaster recovery scenario which is mainly based on vehicles parking along the roadside of streets. It has been shown in [60] that the sum of mean excess delay across all scenarios lies below the $0.8\,\mu$s guard interval. Taps belonging to the maximum delay measured (up to $4.7\,\mu$s depending on the scenario) will only cause a slight amount of interference, because their magnitude, which has been found in [60] to be around $30\,$dB below the strongest tap, should not lead to significant fading behavior.

---

[11]IEEE 802.11p – amendment for wireless access in vehicular environments.

Let us now consider the small-scale fading effects resulting from delay spread. Let us come back to the IMT-advanced UMI NLOS model, where the channel has to be considered as frequency selective as $B_c \approx \frac{1}{5\Delta\tau} \approx 1.5$ MHz $< B_{\text{sig}} = 20$ MHz (see Eq. (3-6)). Thus some of the subcarriers are always likely to experience considerable attenuation due to frequency selective fading which has to be taken into account by an appropriate small-scale fading margin. For practical implementations, the channel can be considered as frequency flat on each of the single subcarriers. The guard interval which is used in IEEE 802.11a (cyclic prefix) is a copy of the last part of the OFDM symbol, which is transmitted before the so-called effective part of the symbol. Thus the effective part of the received signal can be interpreted as the cyclic convolution of the transmitted OFDM symbol with the channel impulse response [61]. The motivation for forming a cyclic convolution in the OFDM signals is to avoid interference of the individual multipath components of the transmission of an individual symbol and to simplify the channel estimation and equalization. This allows for a one-tap equalizer for each of the subcarrier frequencies as it is standard in most OFDM receivers.

The coherence bandwidth is a measure of the maximum frequency difference for which signals are still strongly correlated. The coherence bandwidth B_c can be calculated by Eq. (3-6) (see also [59, p. 163]).

$$B_c = \frac{1}{5\Delta\tau} \tag{3-6}$$

However, it should be noted that an exact relationship between coherence bandwidth and rms delay spread does not exist and Eq. (3-6) is related to the definition that the coherence bandwidth is defined as the bandwidth over which the frequency correlation function is above 0.5. This results in a coherence bandwidth of $B_c = 1.54$ MHz using $\Delta\tau = 130$ ns and shows, that within the wireless bandwidth of the DRS of 20 MHz subcarriers with significant differences in attenuation and phase might be found. Some of the subcarriers may therefore suffer deep fading dips from destructive interference of multi-path components which cause a severe drop in the signal-to-noise ratio on the specific subcarriers. Thus we can confirm that frequency-selective fading in the wireless channel for the DRS under consideration cannot be neglected and has to be taken into account by an appropriate small-scale fading margin.

Before calculating this small-scale fading margin, let us consider the time-selective behavior of the channel. Relative motion between sender and receiver or movement of objects in the channel cause Doppler spread. Doppler spread is of particular relevance for the microscopic multipath components resulting from scatterers in the vicinity of sender and receiver which show only negligible delay spread. Interference of these microscopic multipath components due to their relative frequency shifts leads to strong variations of the received field strength over time periods inversely proportional to the Doppler spread. As a measure of the time interval over which only small changes in field strength at the receiver do occur can be considered the so called coherence time.

The maximum Doppler frequency f_{d_max} can be calculated as: [59]

$$f_{d\_\text{max}} = \frac{v\, f_c}{c} \tag{3-7}$$

where v is the maximum velocity between sender and receiver in m/s, f_c the carrier frequency and c the propagation velocity of the transmitted signal in free space. For OFDM systems Doppler spread may have strongly negative effects on system performance as it leads to Inter Carrier Interference (ICI). ICI only can be neglected if the Doppler spread is much smaller than the subcarrier spacing or, equivalently, coherence time is much longer than the OFDM symbol duration.

Evaluating the expected Doppler spread and corresponding coherence time for a IEEE 802.11 OFDM based DRS we find the following: The velocities occurring in the aftermath of a disaster are assumed to be small as only the users may walk or drive at low velocities through the destroyed area and no fast changes in the environment are expected. Assuming a velocity of 3 km/h the maximum Doppler frequency f_{d_max} with a carrier frequency of 5.5 GHz can be calculated to be 15.28 Hz which is much smaller than the subcarrier frequency spacing of 312.5 kHz for the OFDM signal [35, Tab. 17-5].[12]

The coherence time T_c which is inversely proportional to the Doppler spread can be calculated as rule of thumb by: [59, Eq. 4.40c]

$$T_c = \frac{0.423}{f_{d\_max}} \tag{3-8}$$

The coherence time T_c for $f_{d\_max} = 15.28$ Hz is thus 27.7 ms which is much longer than the symbol time $T_{sym} = 4\,\mu s$. This even holds for a typical VoIP packet with size of 300 byte and duration 424 μs assuming BPSK modulation (see Eq. (3-27) in Section 3.2.2.1).[13]

For the scenarios considered in this work with low packet sizes and low velocities the effects of a time-varying channel do not have to be considered. This holds also for a low number of retransmission when considering the IEEE 802.11 MAC whereby the channel conditions remain the same during this time period (see Fig. 3.12 and Eq. (3-28)) and a retransmission doesn't benefit from changing channel conditions (e. g. through time diversity).

---

[12] In contrast, the results obtained by the measurements in [60] showed Doppler spreads with maximum values of 978 Hz because scenarios with cars moving at high velocities have been considered.

[13] As ICI due to Doppler shift is an important issue for OFDM systems we will have a closer look on the effects to be expected for the IEEE 802.11 based OFDM DRS considered here. It has to be noted that even a small Doppler shift can result in a large ICI. The reason is the following: In a perfect OFDM system each subcarrier is in the spectral nulls of all other subcarriers. However, the slope of the $\sin(f)/f$ shape of the signal in the frequency domain is large near its zeros and therefore may lead to a significant ICI if orthogonality is not respected. Assuming pure LOS conditions the channel Doppler shift is given by a single frequency offset and therefore in principle might be compensated by synchronisation based on the frame preamble. For low velocities in the range of 3 km/h as in the case of a DRS Doppler shift is very small when compared to the subcarrier spacing and therefore ICI can be neglected. For high velocities maximum Doppler shift cannot be neglected any more when compared to the subcarrier spacing (for example a velocity of 100 km/h leads to a Doppler frequency of $f_{d\_max} = 509$ Hz) and thus ICI becomes important. This holds even though the coherence time in general still is longer than the symbol time and in the range of frame duration. For example a coherence time of $T_c = 830.6\,\mu s$ can be calculated from a Doppler frequency of $f_{d\_max} = 509$ Hz. This shows that a large Doppler shift can cause ICI for high packet sizes even under pure LOS conditions. Synchronisation based on the frame preamble in this case doesn't work any more as the Doppler induced time dependent phase shift varies too much from frame preamble to frame preamble.
On the other hand, when assuming NLOS channel conditions and significantly different Doppler shifts for the multipath components the compensation by synchronization is no longer possible and ICI occurs, as well.

In conclusion, assuming that ISI due to delay spread is compensated for the DRS by the guard interval of $0.8\,\mu s$, small-scale fading is essentially due to frequency selective fading. Frequency selective fading thereby can also be understood as independent time selective fading on subcarriers spaced apart by frequency offsets larger than the coherence bandwidth[14], where each of the subcarriers experiences a frequency flat-fading. Thus small-scale fading can be modeled for the case of a pure NLOS scenario by a Rayleigh distribution. In order to account for the loss in SNR due to fading dips which are responsible for symbol errors in subcarriers an appropriate margin for system level simulations can be calculated as follows. Under small-scale fading conditions the field strength at the receiver exceeds a given minimum value $E_{\min}$ not in every situation. The probability of finding a field strength at the receiver with values below $E_{\min}$ is called the outage probability P_{out} and can be calculated by the cumulative density function (CDF) of the field strength. We note that the cdf describes the probability that a certain field strength is not exceeded. For Rayleigh fading, i. e. pure NLOS conditions in a time varying channel $P_{\text{out\_Rayleigh}}$ is given by: [58]

$$P_{\text{out\_Rayleigh}} = \text{cdf}(E_{\min}) = 1 - e^{\left(-\frac{E_{\min}^2}{2\sigma^2}\right)} := 1 - e^{(-1/M_{\text{Rayleigh}})} \tag{3-9}$$

where we have substituted $2\sigma^2/E_{\min}^2$ by the fading margin M_{Rayleigh}. Thereby $E_{\min}$ denotes the mean value of the field strength at the receiver which corresponds to a given outage probability $P_{\text{out\_Rayleigh}}$ and σ^2 denotes its variance. From Eq. (3-9) the fading margin $M_{\text{Rayleigh}_{\text{dB}}}$ in dB can be calculated: [58]

$$M_{\text{Rayleigh}_{\text{dB}}} = -10\log_{10}(-\ln(1 - P_{\text{out\_Rayleigh}})) \tag{3-10}$$

Inserting $P_{\text{out\_Rayleigh}} = 0.1$ for 90% coverage into Eq. (3-10) we get:

$$M_{\text{Rayleigh}_{\text{dB}}} = 9.8\,\text{dB} \tag{3-11}$$

Multipath fading for LOS conditions can be modeled in form of a Rice distribution [58] to account for the effect of a LOS peak. The ratio $A_r^2/(2\sigma^2)$ of the power in the LOS component A_r^2 to the power in the diffuse component $2\sigma^2$ is called the Rice factor K_r. The outage probability which can be expressed by the Rician cdf is shown in Eq. (3-12). The Rician K_r-factor in dB is given by the path loss model, e. g. shown in [54, Tab. A1-7]. $E_{\min}$ denotes the mean value of the field strength for the diffuse component at the receiver which corresponds to a given outage probability $P_{\text{out\_Rice}}$ and σ^2 denotes its variance.

$$P_{\text{out\_Rice}} = \text{cdf}(E_{\min}) = \int_0^{E_{\min}} \frac{E}{\sigma^2} e^{\left[-\frac{E^2 + A_r^2}{2\sigma^2}\right]} I_0\left(\frac{EA_r}{\sigma^2}\right) \, dE \quad 0 \leq E < \infty \tag{3-12}$$

$I_n(\cdot)$ is the modified Bessel function of the first kind with order n. Solving Eq. (3-12) for $E_{\min}$ the Rician fading margin $M_{\text{Rice}_{\text{dB}}}$ can be calculated by: [58]

---

[14]It should be noted that this reasoning corresponds to standard modeling of mobile channels by tapped delay line types of models using independent Rayleigh or Rice statistics for each of the taps that are modeling the macroscopic multipaths.

$$M_{\mathrm{Rice_{dB}}} = 10\log_{10}\left(\frac{2\sigma^2(1+K_{\mathrm{r}})}{E_{\mathrm{min}}^2}\right) \tag{3-13}$$

For $P_{\mathrm{out\_Rice}} = 0.1$ a multipath fading margin of $M_{\mathrm{Rice_{dB}}} = 7.0\,\mathrm{dB}$ can be calculated for the RMa LOS model [54] with $\sigma = 4\,\mathrm{dB}$ and mean value $A_r = 7\,\mathrm{dB}$.

An additional gain on coded bits is given by interleaving as used by IEEE 802.11 when considering frequency selective fading. This gain, however, is not considered here, and the transmission ranges calculated in the following section thus can be considered as lower bounds, i.e. as worst case values.

The calculated margins for shadow fading and small-scale fading can be used for system level considerations and the calculation of link budgets.

3.1.1.3 Link budget and transmission ranges

Table 3.2 shows a simplified link budget for the calculation of the allowed propagation loss at a carrier frequency of 5.5 GHz assuming an IEEE 802.11a/n communication link with 20 MHz bandwidth.

Table 3.2: Simplified Link Budget for ITU IMT-advanced UMi model at 5.5 GHz.

Model		UMi NLOS	UMi NLOS
	Coverage in %	90	99
Tx characteristics:	Tx power in dBm	20	20
	Tx Antenna gain in dBi	5	5
	Transmitter EIRP in dBm	**25**	**25**
Rx characteristics:	Rx Sensitivity in dBm	-92	-92
	Rx Antenna gain in dBi	5	5
	Total Receiver gain in dB	**97**	**97**
Margins:	Shadowing in dB	-5.1	-9.3
	Multipath Fading in dB	-9.8	-20
	Total Margin in dB	**-14.9**	**-29.3**
Results:	**Allowed prop. Loss in dB**	**107.1**	**92.7**
	Transmission Range in m	**60**	**25**

The minimum receiver sensitivity defined in IEEE 802.11 [35, Tab. 17-18] is -82 dBm for Binary Phase-Shift Keying (BPSK) modulation and coding rate 1/2 (data rate 6 Mbit/s). For the calculation a common value of -92 dBm is chosen, reflecting typical performance values of actual off-the-shelf WLAN devices. Typical omnidirectional Wi-Fi antennas are available with a gain between 2 and 9 dBi. Wi-Fi SMD antennas or patch antennas which are used in smartphones will of course have a lower antenna gain. An additional gain for antenna diversity which accounts for the possibility for a receiver to use several receive antennas to exploit multi-path propagation by appropriate signal-combining at the receiver can be included as well. For the calculation of the link budget a typical value for the Rx and Tx antenna gains which is set to 5 dBi is used in Tab. 3.2. This leads to a Transmitter Equivalent Isotropically Radiated Power (EIRP) of 25 dBm and total receiver gain of 97 dB. Thereby the maximum allowed EIRP in the frequency range of 5470 MHz to 5725 MHz according to [62] is 30 dBm.

After subtracting the margins (calculated according to [54] and as detailed in the previous section) we get an allowed path loss of 107.1 dB for 90% coverage and 92.7 dB for 99% coverage for the UMi model. This leads, as calculated in Tab. 3.2, to a transmission range of 60 m for the UMi model whereas for 99% coverage we get 25 m. The shadow fading margin has been calculated for $\sigma_F = 4$ dB as provided by the UMi model. When using a shadow fading margin $\sigma_F = 7$ dB as proposed by [57] for STA to STA links the coverage is slightly reduced to 83% for achieving a transmission range of 60 m. Let us emphasize that these values have to be considered as lower bounds, i.e. as worst case values for the transmission ranges, as it can be assumed that most devices will be equipped with at least two antennas which are operating at least in a receive diversity mode. Thus some additional gain due to receive diversity could be assumed to be always present, but which is not considered within the link budget calculation presented in Tab. 3.2. Additionally, the AP to AP as well as the AP to STA links will have to be considered at $\sigma_F = 4$ dB even according to [57].

The calculated values reflect the coverage of a single communication link. The coverage of the whole WMN, however, is dependent on the available number of nodes and connectivity of these nodes which will be further analyzed in the following section. For system level simulations throughout this work a common transmission range of 60 m will be used for each station which seems a reasonable value according to the arguments given throughout this section.

3.1.2 Connectivity

One of the most fundamental properties of communication networks is connectivity. A communication network is said to be connected if there is a path from any node to any other node in the network [63]. Several concepts, models, tools and methodologies exist for the analysis of connectivity which are summarized e.g. in [63]. Two of the major connection models which are used to describe the establishment of communication links between nodes and used in this work are the unit disk connection model and the SINR connection model.

3.1.2.1 Connection Models

Unit Disk Connection Model

In the unit disk connection model (see Eq. (3-14)), a link between a pair of nodes (i and j) is assumed to be possible if their Euclidean distance is smaller than or equal to a given threshold r which is also denoted as the transmission range in a communication network. [63]

$$x_{ij} = \|x_i - x_j\| \leq r \tag{3-14}$$

where x_i and x_j are the coordinate vectors (e.g. $x_i = (x_{i1}, x_{i2})$ and $x_j = (x_{j1}, x_{j2})$ in two dimensions) of nodes i and j respectively and $\| \cdot \|$ denotes the Euclidean norm[15]. The unit disk connection model captures only the essential geometric aspects in a communication network and thus provides a very simple connection model that may help to simplify the analysis. One of the limitations is that this model doesn't include the real antenna pattern. Another limitation is that fluctuations of the mean received power, e.g. caused by shadowing of large objects, are not considered. Also the model assumes independence of the individual transmission from other nodes in the network and thus does not consider any interference.

A somewhat more realistic model is the log-normal connection model which is discussed in the next section.

Log-Normal Connection Model

The log-normal connection model [63] describes the connection in terms of the received signal strength at node j in dBm, denoted by $\mathcal{P}_{\text{Rx}}(x_i \to x_j)$ which is the difference of the transmission power $\mathcal{P}_{\text{Tx}}$ of node i in dBm and path loss $\text{PL}(x_i, x_j)$ in dB. In order to account for the impact of shadowing caused by obstacles affecting the wave propagation which is referred as shadow fading a random component, denoted by Z (see Eq. (3-15)), is included.

$$\mathcal{P}_{\text{Rx}}(x_i \to x_j) = \mathcal{P}_{\text{Tx}} - \text{PL}(x_i, x_j) + Z \tag{3-15}$$

When the received signal strength $\mathcal{P}_{\text{Rx}}$ at the receiving node j is greater than or equal to a given threshold $\mathcal{P}_{\text{th}}$, it is said that node i and node j are directly connected. The path loss in the log-normal connection model is assumed to follow a logarithmic model as shown e.g. in Tab. 3.1 for the ITU IMT-advanced channel models. The general path loss can be specified as follows:

$$\text{PL}(x_i, x_j) = \text{PL}_0(d_0) + 10\alpha \log_{10} \frac{x_{ij}}{d_0} \tag{3-16}$$

$\text{PL}_0(d_0)$ is the reference path loss in dB at a reference distance d_0 and x_{ij} the Euclidean distance between nodes i and j. The so-called path loss exponent α indicates the rate at which the received signal strength decreases with distance. The reference path loss $\text{PL}_0(d_0)$ is calculated

---

[15]In two dimensions the Euclidean distance for 2 points $x_i = (x_{i1}, x_{i2})$ and $x_j = (x_{j1}, x_{j2})$ is calculated as follows $\|x_i - x_j\| = \sqrt{(x_{i1} - x_{j1})^2 + (x_{i2} - x_{j2})^2}$.

either by using the free space Friis equation [58] or is obtained through field measurements at distance d_0 as used e. g. in Tab. 3.1 for the IMT-advanced channel models. The value of the path loss exponent α depends on the environment and terrain structure and can vary between 2 in free space and 6 in heavily built urban areas [59]. Using Eq. (3-16) we can rewrite Eq. (3-15) as:

$$\mathcal{P}_{\mathrm{Rx}}(x_i \to x_j) = \mathcal{P}_{\mathrm{Tx}} - \mathrm{PL}_0(d_0) - 10\alpha \log_{10} \frac{x_{ij}}{d_0} + Z_\sigma \tag{3-17}$$

where Z_σ is a zero-mean Gaussian random variable (in dB) with standard deviation σ. The value of σ is dependent on the scenario and propagation model as shown in Section 3.1.1.2. Node j is directly connected to node i if and only if the received signal strength $\mathcal{P}_{\mathrm{Rx}}$ is greater or equal than the threshold signal strength $\mathcal{P}_{\mathrm{th}}$, i. e. $\mathcal{P}_{\mathrm{Rx}}(x_i \to x_j) \geq \mathcal{P}_{\mathrm{th}}$. The probability that node j is directly connected to node i, is denoted by $P_r(x_{ij})$ [63]. $P_r(x_{ij})$ is calculated as:

$$P_r(x_{ij}) = \int_{10\alpha \log_{10} \frac{x_{ij}}{r_0}}^{\infty} \frac{1}{\sqrt{2\pi}\sigma} e^{-\frac{z^2}{2\sigma^2}} \, \mathrm{d}z \tag{3-18}$$

where

$$r_0 = d_0 \, 10^{\frac{\mathcal{P}_{\mathrm{Tx}} - \mathrm{PL}_0(d_0) - \mathcal{P}_{\mathrm{th}}}{10\alpha}} \tag{3-19}$$

can be interpreted as the transmission range in the absence of shadowing [63].

If $\sigma = 0$ the log-normal connection model reduces to a unit disk connection model with a transmission range equal to r_0. Assumming symetrical path loss and unidirectional links the previous equations also hold if node i is the receiving node and j the sender.

SINR Connection Model

In the previous models the existence of a connection between a pair of nodes and the existence of a connection between another distinct pair of nodes are independent which simplifies the analysis of the network. The assumption of independence might be valid in a wireless network with a small to medium amount of traffic load where the impact of interference on connections is often small and negligible.

The Signal-to-Interference-plus-Noise-Ratio (SINR) connection model has been widely used to capture the impact of interference [63]. The existence of a directed link between a pair of nodes is determined by the strength of the received signal from the desired transmitter, the interference caused by other concurrent transmissions and the background noise. Let $\mathcal{P}_{\mathrm{Tx}}(i)$ be the transmission power of node i in linear units and let x_k; $k \in \Gamma$, be the location of node k, where Γ represents the set of indices of all nodes in the network. Node j then can successfully receive the transmitted signal from node i if and only if the SINR at x_j, denoted by $\mathrm{SINR}(x_i \to x_j)$, is above a prescribed threshold β. The Signal-to-Interference-plus-Noise-Ratio (SINR) connection model according to [63] is:

$$\mathrm{SINR}(x_i \to x_j) = \frac{\mathcal{P}_{\mathrm{Tx}}(i) \, \mathrm{PL}\sigma(x_i, x_j)}{N_f + \gamma \sum_{k \in T_i} \mathcal{P}_{\mathrm{Tx}}(k) \, \mathrm{PL}\sigma(x_k, x_j)} \geq \beta \tag{3-20}$$

where $\mathrm{PL}\sigma(x_i, x_j)$ denotes the power attenuation from node i to node j and includes both the path loss (e. g. second and third term in Eq. (3-17)) and the shadow fading (e. g. last term in Eq. (3-17)). T_i denotes the subset of nodes that are simultaneously transmitting with node i and using the same frequency band and N_f is the background noise power. The coefficient γ weights the impact of interference. Similarly, if node i receives from node j assuming a bidirectional connection Eq. (3-20) is valid by changing the indices. To summarize, in the SINR connection model the existence of a connection from a node to another node not only depends on the transmit power and the power attenuation of this node pair, but also depends on the activity, transmit power and power attenuation of other nodes. [63]

The unit disk connection model is used in this work for system level simulations in order to clarify some of the basic questions which are posed on the DRS, basically the coverage of the system, how many nodes are needed to establish a sufficient network connectivity and the lifetime of that system. The SINR Connection Model will be used in network simulations in order to additionally account for the influence of shadowing and interference.

3.1.2.2 Network Models and Graph Theoretic Tools for Connectivity Analysis

The establishment of links between a pair of nodes is dependent on the location of the nodes. Thereby the nodes in the network are usually modeled by a point process where a node in the network is uniquely represented by a point. It can be distinguished between deterministic point processes and random point processes. In a deterministic point process points are deterministically placed in a specified region similar to the placement of network devices in a specific deployment area. When using a random point process irregular and random aspects of the network can be captured, such as the location of mobile phone users, location of unplanned wireless access points or irregular location of cellular base stations. In view of the unplanned nature of wireless disaster recovery networks, with some mobile users and additional MSTAs distributed in the area, a random point process has to be applied. Some of the most widely used models to describe the location of nodes and detailed in [63] are the Homogeneous Poisson Point Process, the Inhomogeneous Poisson Point Process and the Uniform Point Process. Further information can be found in [63].

In [63] several graph theoretic tools are presented and results are established using these tools. More precisely, those tools which are widely used in studying network connectivity are Continuum Percolation Theory, Theory of Branching Processes and Algebraic Graph Theory. Continuum percolation, which is an extension of discrete percolation, studies point processes where the points are randomly placed in some continuous space. Two widely studied models in continuum percolation are the Poisson Boolean connection model and the Poisson random connection model [63]. In the following we will focus on Continuum Percolation Theory where a number of well-studied models exist, which are based on homogeneous Poisson point processes.

Two important techniques that have been widely used in the theory of continuum percolation are coupling and scaling. Scaling is a technique which can be used to expand or shrink the distances between all pairs of nodes by a common factor such that the original graph and the new graph have the same graphical properties. By using the scaling technique, one can apply results obtained from a network on a unit area to a network on an area of an arbitrary size.

Coupling is used to establish a correspondence between two networks. It exploits the fact, that one network, which has a certain property, which is however difficult to analyze, can be coupled with a second network that has the same property, but is easier to analyze than the other network. In this way, the analysis can often be greatly simplified. An important example for coupling that will be extensively used throughout this work is the comparison of two networks which are essentially the same except that one network is on a unit square and the other is on a unit torus. The torus that is commonly discussed in random geometric graph theory and continuum percolation theory is essentially the same as a square except that the distance between two points on a torus is defined by their toroidal distance [63], instead of the Euclidean distance.

For a unit torus $A_1^{\mathrm{tor}} = [-\frac{1}{2}, \frac{1}{2}]^2$, the toroidal distance between two points x_1, $x_2 \in A_1^{\mathrm{tor}}$ is given by:

$$\|x_i - x_j\|_{\mathrm{tor}} \;\hat{=}\; \min\left\{ \|x_i + z - x_j\| : z \in \mathbb{Z}^2 \right\} \tag{3-21}$$

where $\mathbb{Z}^2$ is the two-dimensional integer set and $\| \cdot \|_{\mathrm{tor}}$ denotes the toroidal distance norm. The use of toroidal distance allows nodes located near the boundary to probabilistically have the same number of connections as nodes located near the center. Therefore, it allows the removal of boundary effects that are present in a square. The consideration of a torus often implies that there is no need to consider special cases occurring near the boundary of the region and that events inside the region do not depend on the particular location inside the region. In this way, one can avoid the intricate analysis of analyzing special situations that occur when nodes are located near the border of a network, which results in a much simpler analysis. Further information can be found in [63].

In this work, first a toroidal metric is used for the basic system considerations. In a second step, the Euclidean metric is used in order to take boundary effects into account within finite square surfaces. Additionally, scalability is investigated when going over to network simulations. When analyzing the performance of the developed Lifetime Enhancement Algorithm (LEA) in Chapter 6, these boundary effects play an important role.

3.1.2.3 Analytical Estimation of the required Node Density for a connected WMN

The existence of isolated nodes in the disaster recovery network is certainly an undesirable characteristic. In a static network an isolated node cannot exchange any information and is therefore useless. In a scenario with low mobility an isolated node which wants to send or receive information has to wait until it moves into the range of another node or until another node passes by which might cause a high message delivery delay. In [64] an analytical expression is derived which allows the determination of the required transmission range r_0 that creates, for a given node density $\rho = N/A$ with N nodes in an area of size A, an almost surely k–connected network. To this end, in [64] the topology of an ad-hoc network is represented as an undirected geometric random graph, denoted as $G(N, K)$ with N nodes and K wireless links. A random geometric graph thereby is a geometric graph in which the N nodes are independently and uniformly distributed in a random manner within a metric space. Furthermore a graph is said

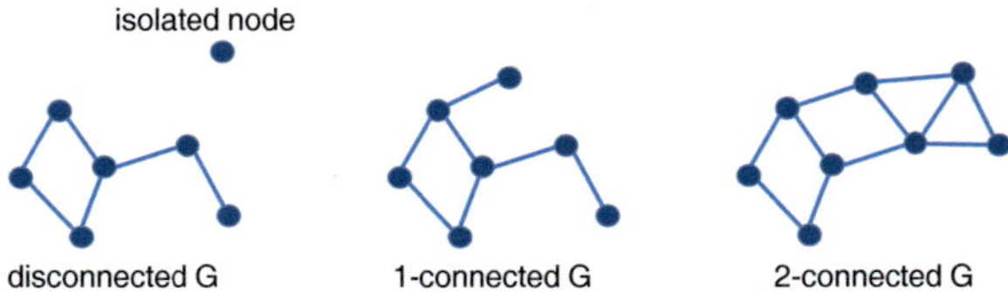

Figure 3.2: Sketch of a disconnected graph G, a 1-connected graph G and a 2-connected graph G.

to be k–connected ($k = 1, 2, \ldots$) if for each pair of nodes there exist at least k mutually independent paths connecting them (see Fig. 3.2), e. g. if $k = 2$ and one connection of a node fails there still exists another connection which connects this node to all other nodes. The maximum value of k for which a connected graph is k–connected is denoted as the connectivity of the graph. Equivalently, if the maximum range r_0 of the nodes is given, it can be calculated how many nodes of this type are needed to cover a certain area in order to get a k–connected network when assuming the unit disk connection model. The degree of a node i, denoted as $d(i)$, is the number of connected neighbors of node i, i. e. its number of links. A node of degree $d = 0$ therefore is isolated, i. e. it has no connected neighbor nodes [64]. The minimum node degree of the whole graph, this is the node degree of the node which has the smallest node degree is denoted by $d_{\min}$. A graph with $d_{\min} > 0$ thus is a graph with no isolated node. As the node degree d is always a positive integer $d_{\min} > 0$ is equivalent to $d_{\min} \geq 1$. In the following we will only use the second notation. One of the fundamental questions for the proposed DRS in this work is to determine the required node density such that, with high probability, no node in the network is isolated (i. e., each node has at least one connected neighbor) or, equivalently, to determine the required minimum range r_0 in order to achieve a fully connected network in the disaster area. To solve this problem, the nearest neighbor method known from analysis of spatial data, e. g. as described in [65, Eq. 8.2.9] has been used in [64]. Thereby a set of N nodes, each node with a transmission range r_0, are randomly and uniformly placed in a large area $A \gg r_0^2 \pi$ following a uniform two dimensional distribution and respecting a mean node density $\rho = N/A$. The distance of a point to its closest neighboring point is denoted as its nearest neighbor distance ξ [64],[65]. In order to apply the theoretical result in [65, Eq. 8.2.9] we have to note, that a homogeneous Poisson Point Process on two dimensions equals – using Cartesian coordinates – a uniform distribution of the random variable that describes the location of the individual nodes. For a homogeneous Poisson point process in two dimensions, the probability density function $f(\xi)$ of the nearest neighbor distance according to [65, Eq. 8.2.9] can be calculated as follows:

$$f(\xi) = 2\,\pi\,\rho\,\xi\,\mathrm{e}^{-\rho\pi\xi^2} \quad \text{for } \xi > 0 \tag{3-22}$$

As stated in the previous section a random node in the disaster recovery network is uniquely represented by a random point in the network. Thus the probability that the distance between a randomly chosen node and its nearest neighboring node is less than or equal to r is: [64]

$$P(\xi \leq r) = \int_0^r f(\xi)\,\mathrm{d}\xi = 1 - \mathrm{e}^{-\rho\pi r^2} \tag{3-23}$$

We can use Eq. (3-23) to analyze the probability to achieve a connected network when assuming the unit disk connection model: Eq. (3-23) thereby denotes the probability that a node i has at least one connected neighbor node when assuming the same transmission range $r = r_0$ for all nodes which in particular holds when using the unit disk connection model. Using the node degree $d(i)$ we can rewrite Eq. (3-23) for all nodes in the considered set of nodes and assuming an identical transmission range r_0 for all nodes as follows:

$$P(d(i) \geq 1) = P(\xi \leq r_0) \qquad \text{for all } i \text{ in } \Gamma \tag{3-24}$$

The probability that a node is isolated, i.e. having a node degree $d(i) = 0$, is therefore: [64]

$$P(d(i) = 0) = P(\xi > r_0) = 1 - P(\xi \leq r_0) = \mathrm{e}^{-\rho\pi r_0^2} \tag{3-25}$$

In order to achieve a network graph G in which none of the N nodes is isolated and assuming statistical independence of the positions of the individual nodes, the probability for this event can be calculated from a Bernoulli distribution as follows and noting that a graph with no isolated nodes can be described by a minimum node degree greater than zero: (see [64])

$$P(d_{\min} \geq 1) = \binom{N}{N} P(d > 0)^N \, P(d = 0)^0 \tag{3-26}$$
$$= (1 - \mathrm{e}^{-\rho\pi r_0^2})^N$$

The expression in Eq. (3-26) thus describes the probability that no node is isolated.

This analytically derived equation assumes a large area $A \gg r_0^2\,\pi$. However, nodes at the border of a small finite area have less connections to other nodes than nodes in the middle. Due to this border effect Eq. (3-26) can be seen as an upper bound on the probability that all nodes are connected. Therefore the actually required node density to achieve full connectivity is always higher when boundary effects are taken into account. Figure 3.3 shows the probability $P(d_{\min} \geq 1)$ that no node is isolated for $A = 1000\,\mathrm{m} \cdot 1000\,\mathrm{m}$ as a function of the number of nodes N in the area A and of the transmission range r_0 being equal for each node. Examining Eq. (3-26) it easily can be found, that for the size A of the considered area approaching infinity an infinite node density is needed in order to ensure that no node is isolated. This essentially is due to the fact that the positions of the nodes are statistically independent.

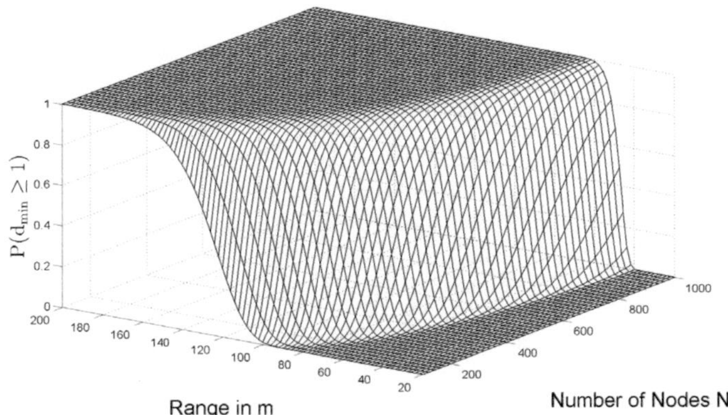

Figure 3.3: Probability that no node in the network is isolated as function of the transmission range r_0 and of the number of nodes N according to Eq. (3-26), $A = 1000\,\text{m} \cdot 1000\,\text{m}$.

For a basic analysis of the required node density to successfully establish a DRS Eq. (3-26) is therefore helpful only for a restricted range of sizes A of the considered area. The area A thereby on one hand has to be large when compared to the disk πr_0^2 covered by the transmission range of a single node, on the other hand A has still to be relatively small in order not to predict unreasonably high node densities. For very large areas, however, the criterion of a fully 1-connected network is by far too stringent.

Intuitively it is clear, that – under the assumption that the problem of coverage achieved by the DRS scales properly with the size of the area – for large areas there should exist a constant node density that is sufficient to achieve an acceptable coverage of the area. It is also intuitively clear, that "acceptable coverage" thereby means, that a certain number of isolated nodes has to be accepted.

Therefore a different approach has to be used which focusses on the probability for achieving a sufficient coverage. As already stated, the reason for this is as follows: on a very large area, few unconnected nodes are always tolerable and the cost of connecting all the nodes is just way too high for a huge number of nodes in view of a wireless mesh network in a disaster area. This will be further discussed in Chapter 4.

3.1.3 Lifetime of WMNs and Proposals for their Energy Efficient Realization

An overview of different energy conservation mechanisms for WMNs at network layer, MAC layer and physical layer is given in [66]. Connectivity, the calculation of a critical transmission range as well as the adjustment of power levels that allow maintaining full network connectivity have been investigated in several research papers (e. g. [64], [67], [68], [69] and [70]). In contrast to the focus of existing research, in this work, connectivity should be analyzed with particular emphasis on the lifetime of the system.

The research on optimum transmission radii for packet radio networks with mesh topologies goes back to the middle of the 1970s. [71] already showed in 1978 an equation for network throughput as a function of the average number of connected neighbors for each node, which is also called the average node degree.

Throughput is not the most relevant parameter in this work, but it will be shown, that coverage can be guaranteed to a sufficient level by just respecting a sufficient number of connected neighbors for each station. One of the key factors affecting the capacity of the network thereby is the transmission radius of the nodes. As already shown in the previous section [64] describes an analytical expression that enables the determination of the required range r_0 that creates, for a given node density ρ, an almost surely k–connected network, where we have focussed on the case of a 1-connected network, i. e. on the absence of isolated nodes. A large transmission radius leads to a high probability of a connected network, however, with the drawback of more interference and corresponding channel loss. A small transmission radius causes less interference but higher risk of a non-connected network and thus more hops are needed for packet forwarding which also results in a higher network traffic and therefore reduction of the network capacity. [71] further showed an optimal average number of connected neighbors, i. e. the number of neighbors within the transmission range for each node, of approximately 6 nodes which is also called "the magic number". In [72] this problem was reconsidered showing a different "magic number" between 7 and 8 for different channel access schemes. In [73] different routing strategies were analyzed, showing that the number of connected neighbors should be between 6 and 8 where nodes are distributed in the plane according to a homogeneous two-dimensional Poisson process. In contrast [74] states that the number of nearest neighbors, i. e. the number of connected neighbor nodes, should increase as the number of nodes in the network increases in order to avoid a disconnected network. This is in line with the analytical considerations in [64] where we emphasize again, that a fully connected network is not a reasonable goal for a DRS, where only a sufficient coverage has to be achieved by the DRS in a given area.

More recently, the so-called coverage problem which is mainly addressed in the field of wireless sensor networks has been studied in view of several objectives, assumptions and constraints. The different approaches to solve the coverage problem can be classified according to [75] into deterministic or stochastic sensor deployment as well as deployment with prior knowledge about the location of the sensors. Additionally, homogeneous or heterogeneous sensing areas are considered, as well as additional design constraints such as energy efficiency, minimum number of sensors, that need to be deployed, or network connectivity.

[67] states that every node should transmit with just enough power to guarantee connectivity of the overall network and derives the critical power a node needs in order to ensure connectivity. [68] investigates on the optimal transmission range for wireless ad-hoc networks based on their

energy efficiency and determines a single static optimal energy-efficient transmission range for all nodes. [69] proposes an algorithm to select the routes such that the energy consumption is balanced among the nodes in proportion to their energy reserves. In some research papers (e. g. [68], [70]) it is also stated that the optimal transmission range can be set at the system design stage or pre-deployment phase in order to minimize the total system energy. This can only be true for planned networks such as sensor networks in a specified area, but not in the proposed DRS of this work.

The characteristics of the DRS proposed in this work, especially the fact that only a low amount of traffic has to be carried by the system, led to the conclusion that the methods discussed in [67], [68], [69] and [70] (transmission range power adjustments or load balancing) do not beneficially enhance the lifetime of the system, as power consumption of the nodes in the DRS is not throughput driven, but dominated by the stand-by state energy consumption of the MSTAs. Thus the lifetime of the DRS is mainly based on the stand-by lifetime of each individual node and on the remaining connectivity when nodes run out of power. Any transmission range power balancing or load balancing would therefore not have a major effect on increasing the overall lifetime of the DRS.

The setting of an optimal transmission range at the system design stage or pre-deployment phase as proposed in [68] and [70] obviously is also not possible for the DRS proposed here. In the approach studied in this work the network is assumed to be dense and more nodes than needed to cover the disaster area are present[16]. The transmission radii are assumed to be constant (homogeneous sensing area) and in a first attempt we are not interested in maximum capacity of the network which basically needs to enable a low number of contemporaneous emergency calls. In view of this, some nodes can be shut down and reactivated when needed to extend the lifetime of the whole network. One approach, following the results in [71], [72] and [73], could be to consider the minimum number of connected neighbors for the decision which nodes to shut down while still ensuring connectivity of the network. A side effect thereby is a reduced congestion of the network due to a reduced number of nodes active, leading to a number of neighbors for each node closed to optimum. This approach of keeping the number of active nodes in the network close to optimum will be further analyzed and requires a type of Sleep/Wake algorithm for each individual node.

Sleep/Wake scheduling algorithms have been studied extensively in the context of wireless sensor networks, e. g. in [76] and [77]. One approach to conserve energy and described in [76] is to put the radio to sleep during idle times and wake it up right before message transmission or reception by using an appropriate TDMA protocol. TDMA protocols thereby avoid energy waste due to contention. However, a precise synchronization between the sender and the receiver is required, so that they can wake up at the same time to communicate with each other. These algorithms proposed for wireless sensor networks are not suitable for the DRS

---

[16]This seems to be a reasonable assumption which is also confirmed by field studies on the density of WLAN devices per area [17]. This is further discussed in Section 4.1.

proposed in this work as the DRS requires availability at any time, e. g. for an emergency call. [77] showed another effective approach for energy conservation in wireless sensor networks by scheduling sleep intervals for extraneous nodes while the remaining nodes stay active to provide continuous service. One of their assumptions, however, is that every node knows its accurate location which is obviously not the case in the approach studied in this work.

In [78] a novel routing algorithm for IEEE 802.11-based wireless mesh networks is presented. The algorithm accounts for energy considerations by using as few nodes as possible, which allows switching off the unused nodes and thus to save energy. However, the execution of the routing algorithm itself is done at a centralized location of the mesh network which is called the Centralized Server. The nodes in the network of a DRS as proposed in this thesis have to act individually because a central server is not available. Therefore in this work the development of a distributed algorithm is proposed where the nodes decide individually if they are needed or not without knowledge of the whole network and node locations.

Additionally, the Power Saving Mode (PSM) of the IEEE 802.11 Protocol [35, Section 13.14] can be used to reduce the energy consumption of a device. Three modes (active, light sleep and deep sleep) are available in IEEE 802.11s WMNs. The decision when a node goes in PSM is, however, out of the scope of the standard and needs to be coupled with an appropriate algorithm as will be done in Chapter 5 of this thesis.

3.2 Network Simulation of Wireless Mesh Networks

The setup of complex large-scale wireless mesh networks as DRS in real-time scenarios would be very time-consuming, expensive and in some cases even not possible. Network simulations, however, allow for a relatively fast simulation of complex scenarios with reproducible results and easy change of settings, protocols and environment. The accuracy of the simulation results is always dependent on the accuracy of the simulation models and level of simulation detail which is also a trade-off between the development of accurate, but complex models and simulation time while all aspects of reality can never be included. As will be shown, it is a strong requirement that the used network simulation model needs to incorporate key characteristics of real hardware in order to achieve more reliable results. Those key parameters thereby can be estimated by small testbed measurements under the assumption that scalability to larger networks is given. One example is the estimation of processing delays which are added by each WLAN module during packet forwarding due to software execution time and the scheduling mechanism of tasks in the CPU. Processing delays are rarely modeled in network simulators which may result in incorrect assumptions when analyzing delay-sensitive applications as will be shown throughout this thesis.

An appropriate network simulator has to be chosen in accordance with the needs and adopted criteria of evaluation. General criteria according to [79] are flexibility, availability of models, scalability, graphical facilities, hardware/software considerations, ease of use, output reports, plots, documentation and support. An analysis in [79] of selected simulators in publications (IEEE, Springer, Wiley and Elsevier from 2000 to 2013), related to wireless networks, shows that ns-2 was the most popular simulation tool during that time. Further ns-3, OPNET, OMNeT++, GloMoSim and QualNet are mentioned. ns-2 and ns-3 are both open source (licensed under GNU GPLv2), packet-based, discrete-event network simulators whereby ns-3 is the successor of the successful Network Simulator 2 (ns-2) which since 2011 is no more under active development. OMNeT++ is an extensible, modular, component-based C++ simulation library and framework, primarily for building network simulators licensed under the Academic Public License [80]. It supports the simulation process for wireless mesh networks, however, there is no implementation of the full IEEE 802.11s stack. The current version consists of the implementation of the IEEE 802.11s routing protocol which is based on the implementation in ns-3. OPNET and QualNet are both commercial tools where QualNet is the commercial Version of GloMoSim which is also no more under active development. Finally, it is concluded in [79] that for academic researchers ns-3 would be the best choice. In [81] the performance of different open source network simulators (ns-2, ns-3, OMNeT++ and GloMoSim) has been evaluated in a setup with AODV routing algorithm and varying number of nodes from 400 to 2000. It has been shown that with increasing number of nodes, a linear growth in memory consumption for all simulators can be observed with minor differences. ns-2 has shown the highest computation time which increases rapidly with increasing number of nodes and has therefore a low scalability. In terms of computation time and scalability, ns-3 appeared to be the most efficient one in [81]. From 2006 to 2008 ns-3 was developed based on ns-2, the Yet Another Network Simulator (YANS) project developed by Mathieu Lacage at Inria[17] and the Georgia Tech Network Simulator (GTNetS). It has to be noted that ns-3 is not just an update of ns-2. The core was completely rewritten based on the successful and unsuccessful experiences of ns-2 without backward compatibility.

The Network Simulator 3 (ns-3) has been chosen in this work because of its flexibility, expandability and free use in terms of the GNU GPLv2 license, scalability and finally because of the already available IEEE 802.11s mesh model. Further ns-3 has the possibility to generate pcap[18] traces of the simulated models. This allows to easily debug the results with standard tools such as Wireshark [82] and to compare them with results obtained by real measurements which will be used in Part II of this work. The following section gives an overview of ns-3 and details the required validation of the individual models for this work.

---

[17]Inria - Institut national de recherche en informatique et en automatique, The French National Institute for Research in Computer Science and Automation.

[18]Packet Capture Data file type which contains network packet data, e. g. created during a live network capture.

3.2.1 Overview of ns-3

The network simulator ns-3 [83] is a packet-based, discrete-event network simulator with special focus on Internet-based systems, targeted primarily for research and educational use[19]. ns-3 simulation scripts are basically written in C++. Additionally, support for extensions that allow simulation scripts to be written in Python is given. The simulator consists of two simulation modules (core and network) which are intended to comprise a generic simulation core that can be used by different kinds of networks. The core of the simulator comprises those components that are common across all protocol, hardware and environmental models. Important elements of the generic simulation core are the Simulator class which is responsible for the control of the virtual time and the execution of simulation events and the Scheduler class which manages the event list by creating and scheduling events. Fundamental objects of the network simulator such as Packets, Nodes, NetDevices, Queues and Sockets are implemented in the network module. Figure 3.4 shows an overview of the software organization in ns-3. The mobility module comprises models such as random walk, constant velocity or constant position. The Internet module consist e. g. of IPv4, IPv6 routing protocols, TCP and UDP models. Application consists e. g. of a traffic generator which follows an On/Off pattern which is used in this work. Propagation consists of several propagation models which will be described throughout the following sections. The helper API is used to make ns-3 programs easier to write and read, without taking away the power of the low-level interface [84].

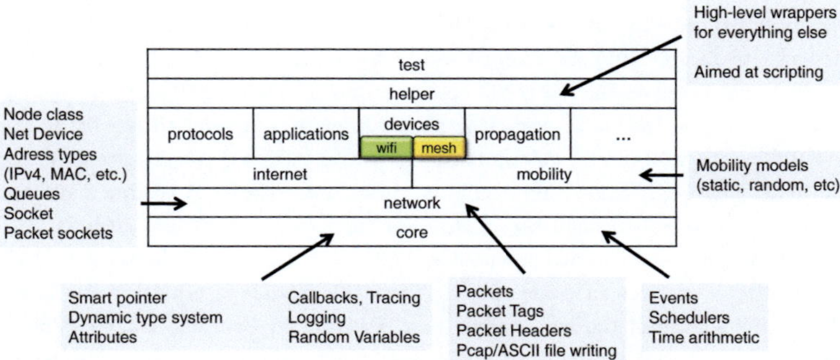

Figure 3.4: ns-3 software organization. [84]

One of the key simulator objects is a node (Fig. 3.5). A node consists of a NetDevice, Protocol stack, Application and Packets. The NetDevice represents a physical interface on a node (such as an Ethernet interface). The basic idea is to mimic the Linux architecture at the boundary between the device-independent sublayer of the network device layer and the IP layer. The

---

[19]ns-3 specific terms, models, objects and message types use exactly the spelling as used in the ns-3 documentation, including upper and lower case spelling.

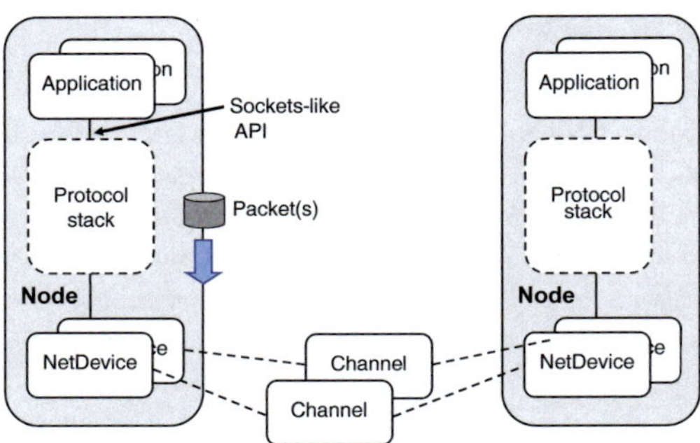

Figure 3.5: ns-3 node object overview. [85]

IPv4 or IPv6 portion of a device is modelled by a separate object on top of the NetDevice, the Protocol stack. Applications are user-defined processes that generate traffic to send across the networks to be simulated. Development of different types of applications that have different traffic patterns is possible. The communication of applications on a node with the node's protocol stack is handled via sockets. Each NetDevice of a node can be associated to a channel through which packets can be exchanged with other nodes.

The ns-3 IEEE 802.11s mesh module [86] was built separately from the existing wifi module and doesn't break the existing IEEE 802.11 MAC model. The mesh module architecture was designed to flexibly support different Wi-Fi based Layer2 mesh solutions. Initially the mesh module was based on the IEEE 802.11s Draft Standard 3.0 (2009) and has been updated during this work (see [23]) with regard to packet formats compliant to the IEEE Standard 802.11-2012[20]. The ns-3 mesh model implements only a subset of the service extensions defined in IEEE 802.11s. It is focusing mainly on those items related to peering and forwarding (routing) of data frames through the mesh. Not supported are features such as MCCA as well as security and power saving mechanisms [83]. An overview of the supported and not supported features is given in Appendix B. The mesh module is using the low layer MAC and PHY functionality of the wifi module (e. g. Distributed Coordination Function (DCF), PHY, Channel) and adds the required upper layer functionality (e. g. MeshWifiInterfaceMac, MeshPointDevice, MeshL2RoutingProtocol). This approach can be illustrated by the following example: The MeshPointDevice uses the "lower" MAC implementation of the WifiNetDevice of the wifi module (see Fig. 3.6).

The WifiNetDevice models a wireless network interface controller based on the IEEE 802.11 standard [83]. The IEEE 802.11 model provided in ns-3 realizes an accurate MAC-level implementation of the IEEE 802.11 specification and a packet-level abstraction of the PHY-layer. Due to this abstraction it is not possible to capture the effects of fast fading or frequency-selective channels as it would be required for a pure physical-layer simulation. However, a

---

[20]These changes are still valid for IEEE 802.11-2016.

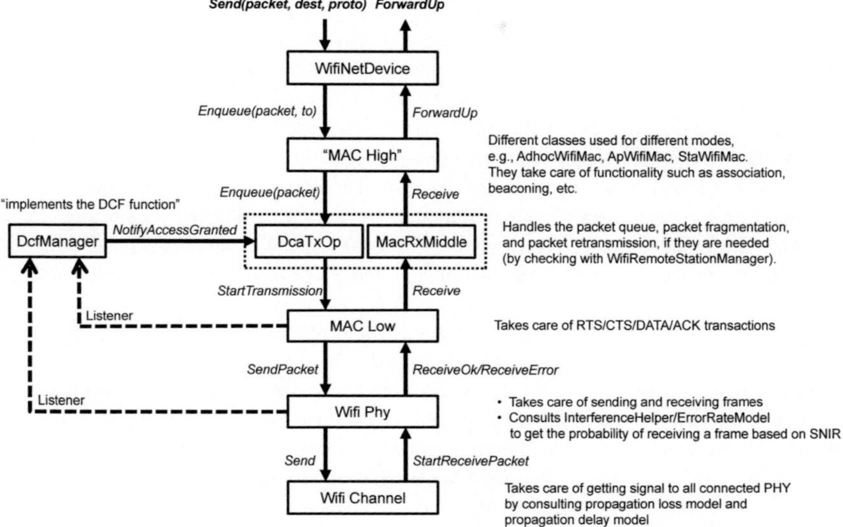

Figure 3.6: WifiNetDevice architecture as shown in [83].

suitable abstraction may be derived that covers all key aspects to properly model the effects of time and frequency selective fading relevant on network level. This abstraction of the physical-layer is detailed in Sec. 3.2.3, the required changes and amendments are detailed in Sec. 7.1. In [87] a physical layer emulator (PhySim-Wifi v1.2 for ns-3.13, 2012) for OFDM-based IEEE 802.11 communications for ns-3 has been developed. However, the increased accuracy of the physical layer model by operating on complex time samples instead of working on frames as a whole comes with the drawback of a huge increase in simulation time by a factor between 330 and ~160.000 when using the physical layer emulation, depending on the frame size and whether a path loss only, a slow- or a fast-fading channel is simulated [87]. This extension is not part of the current ns-3 release and would also be not suitable for complex simulations with a large number of nodes as used in this work.

Additionally, Section 3.1.1.2 showed that for the scenarios considered in this work with low velocities the effects of a time-varying channel do not have to be considered explicitly, but it is sufficient to statistically cover these effects by appropriate margins in the path loss and the SINR connection model. Frequency selective channels caused by delay spread are covered on one hand by the small scale fading margin and on the other hand ISI due to delay spread is assumed to be entirely avoided by the sufficiently large cyclic prefix of $0.8\mu s$ (see Section 3.1.1.2).

The WifiNetDevice architecture (Fig. 3.6) is modular and provides three sublayers of models: the PHY layer models, the MAC low models and MAC high models. The MAC low models include functions such as medium access (DCF and EDCA), RTS/CTS[21]/DATA and ACK transactions. The lower-level MAC is further subdivided into a MAC low and MAC middle sublayer, with channel access grouped into the MAC middle. The MAC high models implement non-time-critical processes in wifi such as the MAC-level beacon generation, probing and association state machines as well as a set of rate control algorithms. Three MAC high models are specified in the wifi model for the configuration as Access Point, non-AP Station (Client) and STA in an Independent Basic Service Set referred as ad-hoc network. The MeshWifiInterfaceMac for the configuration of a Mesh Station, however, is defined in the mesh model. These four MAC high models share a common parent in ns3::RegularWifiMac, which exposes also an attribute *QosSupported* which allows the configuration of IEEE 802.11e QoS support. [83]

The physical layer models are mainly responsible for modeling the reception of packets and for tracking the energy consumption. Each packet received is probabilistically evaluated for successful or failed reception. The probability depends on the modulation, on the signal to noise (and interference) ratio of the packet and on the state of the physical layer. The Yans-WifiPhy model has been the only physical layer model until recently and is implemented in the YansWifiPhy class which is described in [88]. One of the limitations of the current physical layer implementation is, that the model doesn't support the coexistence of signals from different technologies. To overcome this limitation the SpectrumWifiPhy class has been newly introduced which is still under development. Also the frame capture effect is currently not included in the YansWifiPhy which means that the transceiver is unable to tune on a stronger signal while receiving a weak signal. This will be discussed in more detail in Section 3.2.3.
The YansWifiPhy model specifies two models for the calculation of the bit error rates and corresponding packet error rates for orthogonal frequency division multiplexing, the YANS [88] and NIST[22] [89] error rate models. In [89] both models are validated by empirical hardware measurements and the NIST model is recommended for the calculations because of the overly optimistic results in the YANS model. See Section 3.2.3 for further validation results obtained in this work.

The ns-3 wifi model is on ongoing development, where due to the open source nature still some inaccuracies might be present. Not much research work addressing the validation of the ns-3 wifi network simulation models are present. [89] focusses on the validation of the physical layer and channel models, however, it will be shown in the following sections that still some inaccuracies are present. In order to improve the credibility of the network simulation [90] validates the IEEE 802.11 MAC layer model by means of measurements in the EXTREME testbed. It is also stated in [90] that, while in general there is a good qualitative agreement between simulator and testbed, in several cases there were noticeable quantitative differences. The reason might not only be due to an inaccurate ns-3 implementation, but even more likely due to a rather non-standard behavior of some commercial cards which are aimed on performance enhancement [90].

---

[21]Request to Send (RTS) / Clear to Send (CTS) is the optional mechanism used by the IEEE 802.11 wireless networking protocol to reduce frame collisions introduced by the hidden node problem.

[22]National Institute of Standards and Technology (NIST).

The following sections details the validation and – where necessary – the required improvements of the fundamental functionalities provided by ns-3 mesh, ns-3 wifi MAC and ns-3 wifi PHY models which are key for a reliable evaluation of the appropriate network performance of the proposed DRS in this work. More advanced topics such as the modeling of processing delays are detailed in Chapter 7. The results of the following validation and recommended changes based on ns-3.24 have also been published in [18], [21] and [22].

3.2.2 Validation of ns-3 wifi and mesh MAC

The current simulation model in ns-3 lacks the modeling of appropriate processing delays which are added by each WLAN module due to software execution time and the scheduling mechanism of tasks in the CPU. In ns-3 the actual packet processing time of the Mesh Stations (MSTAs) is exactly zero and the Hybrid Wireless Mesh Protocol is highly sensitive to this fact causing an unrealistic behavior or even doesn't work at all (see bug reports 912, 1465 and 1605 [91]). One rather striking explanation of this sensitivity can be understood considering the following scenario: One MSTA (n_0) is sending a broadcast Path Request (PREQ) message. Two or more MSTAs (n_1, n_2, ...), which are located at exactly the same distance to the sending MSTA are receiving this packet (Fig. 3.7a).

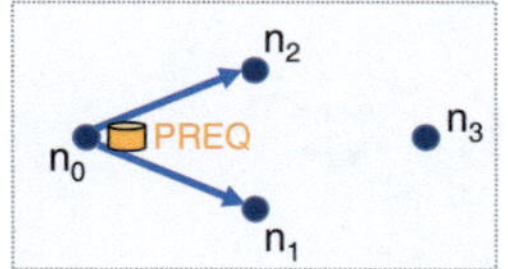

(a) MSTAs n_1 and n_2 receiving PREQ packet.

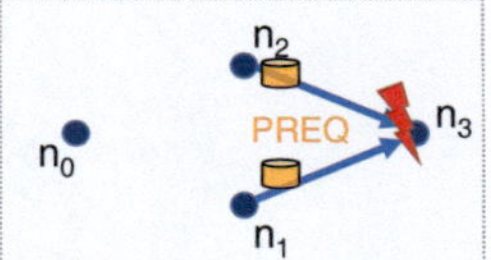

(b) PREQ packets colliding at MSTA n_3.

Figure 3.7: ns-3 mesh path establishment problem.

The received packet goes immediately through the stack of all the MSTAs (n_1, n_2, ...) and to their outgoing wifi-queues. All received packets are forwarded immediately at the same simulator time. Backoff is not calculated in this scenario in the current ns-3 implementation, because the simulator assumes that there was no previous concurrency in the network. It has to be noted that this is not correct: The MSTAs are receiving the Path Request (PREQ) messages and the medium thus previously was not IDLE. In this implementation of ns-3 packets are thus colliding (see Fig. 3.7b) and because of the broadcast mechanism (broadcast packets are not acknowledged) packet loss is not recognized. In this specific case a path can never be established by the path selection protocol. The problem is rather a not entirely correct implementation of the MAC as will be detailed in a moment than not considering processing delays. However, the inclusion of processing delays leads to a more correct behavior of the MAC and would have solved this issue, as well.

The inaccurate MAC implementation leading to the problem described above can be analyzed as follows: According to the DCF access procedure described in IEEE 802.11 [35, Chapter 10] which is the base for MCF a MSTA achieves immediate access to the medium when the MSTA determines that the medium is idle for greater than or equal to a DCF InterFrame Space (DIFS) or Arbitration InterFrame Space (AIFS) period in case of EDCA (see also Appendix A). However, this is not the case in the described example except for the first station if the medium previously was IDLE. Except for the first station the packet thus should not go immediately to the outgoing wifi-queue without generation of a random backoff period. Instead backoff should be performed which has also been discussed in ns-3 bug report 555. Winlab Rutgers [92] proposed a patch for ns-3.16 in order to change the DCF manager to fix this issue. This patch has been included in ns-3.26, but has been used even with an earlier version of the simulator throughout this work.

3.2.2.1 Analysis of the mesh functionality using a linear chain topology

The behavior of the ns-3 wifi MAC protocol in conjunction with the mesh model will be analyzed in a simple forwarding scenario by setting the Mesh Stations in a linear chain topology with up to five MSTAs (see Fig. 3.8).

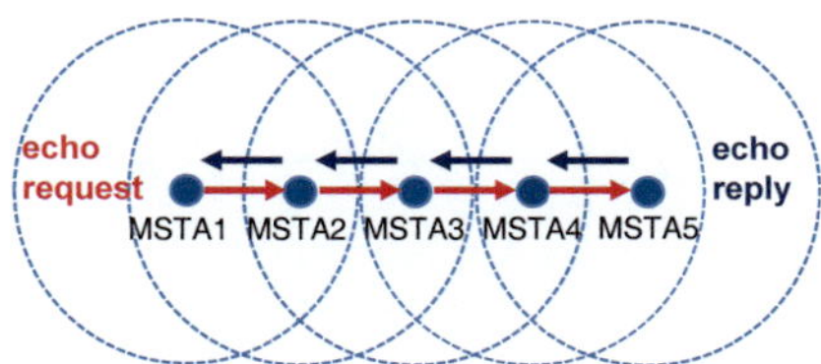

Figure 3.8: Schematic of the Mesh Station setup as linear chain topology.

Every node has connection to its direct neighbors one hop apart using a propagation model with fixed range namely the ns-3 Range-Propagation-Loss-Model. This model corresponds to an implementation of the unit disk connection model discussed in Section 3.1.2.1. It has to be recognized that every node which is found 2 hops away is a hidden node to the respective sending node which may cause collision of packets when transmissions are overlapping. The simulations have been carried out for chain topologies with 2, 3, 4 and 5 MSTAs.

The time introduced by each additional hop can be estimated by performing a ping from the first to the last MSTA of the chain. Thereby ping is operating by sending an Internet Control Message Protocol (ICMP) echo request to the target MSTA and waiting for an ICMP echo reply. By appropriately choosing the transmission range which is equal for all MSTAs as well as the positions of the MSTAs, each MSTA establishes a Peer Link only to its direct neighbors. The reactive on-demand mode of the hybrid wireless mesh protocol (HWMP) is used in order to establish a route between MSTA1 and the rightmost MSTA of the chain and the Round-Trip-Time (RTT) is determined for each configuration by the used ping.

The ping is performed with an interval of 20 ms and 222 bytes of payload. Table 3.3 shows the protocol and header lengths of the packets for the ns-3 simulation. The payload in the ns-3 simulation has been adjusted to meet the packet size of the testbed measurements carried out in Chapter 7. The duration t_p of the packets can be calculated by Eq. (3-27): N_{DBPS} is the number of data bits per OFDM symbol. b is the size of the Physical Layer Convergence Procedure (PLCP) Service Data Unit (PSDU) in bytes. [35, Fig. 17-1] shows the structure of a physical layer frame which consists of the PHY preamble, the OFDM PHY header, PSDU, tail bits and pad bits. The PHY header consist of a single OFDM symbol, denoted as SIGNAL, and the SERVICE field of 16 bits. After the PSDU 6 zero tail bits are appended. Additionally, the timing parameters according to [35, Tab. 17-5] associated with the OFDM PHY are the PHY preamble duration of 16 μs and the OFDM Symbol interval T_{sym} which is 4 μs. Assuming Binary Phase-Shift Keying (BPSK) with coderate 1/2 (datarate 6 MBits/s), N_{DBPS} equals 24 bits. With a packet size of 300 bytes this finally leads to a packet duration of $t_p = 424\,\mu$s. The duration of an ACK packet t_{ack} with a length of 16 bytes leads to a duration of 44 μs.

$$t_p = \text{PHY preamble} + \text{SIGNAL} + \left\lceil \frac{\text{SERVICE} + b \cdot 8\frac{\text{bit}}{\text{byte}} + \text{tail bits}}{N_{\text{DBPS}}} \right\rceil T_{\text{sym}}$$

$$= 16\mu s + 4\mu s + \left\lceil \frac{16\text{bit} + b \cdot 8\frac{\text{bit}}{\text{byte}} + 6\text{bit}}{N_{\text{DBPS}}} \right\rceil 4\frac{\mu s}{\text{Sym}}$$

$$(3\text{-}27)$$

Table 3.3: Packet compilation for the 222 bytes payload ICMP packet used in the linear chain topology simulation (see Fig. 3.8).

Protocol	Length (Bytes)
IEEE 802.11 QoS Data	42
Logical-Link Control	8
IPv4	20
ICMP	8
Payload	222
Packetsize b	300

3.2.2.2 Simulation results

Figure 3.9 shows the histogram of round-trip-times for the configurations with 2, 3, 4 and 5 MSTAs. In each configuration 100 simulation runs have been performed with different seeds. In each simulation run with a duration of the ping of 60 seconds 3000 ICMP packets have been sent resulting in 300.000 packets per configuration. The configuration with 2 MSTAs shows a single peak at 0.951 ms. This time can be calculated as shown in Fig. 3.10.

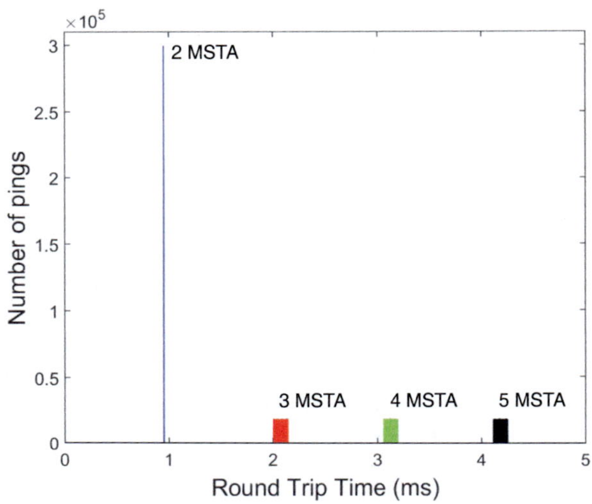

Figure 3.9: Histogram of round-trip-times for the network simulation setups according to Fig. 3.8 with chains of 2, 3, 4 and 5 MSTAs.

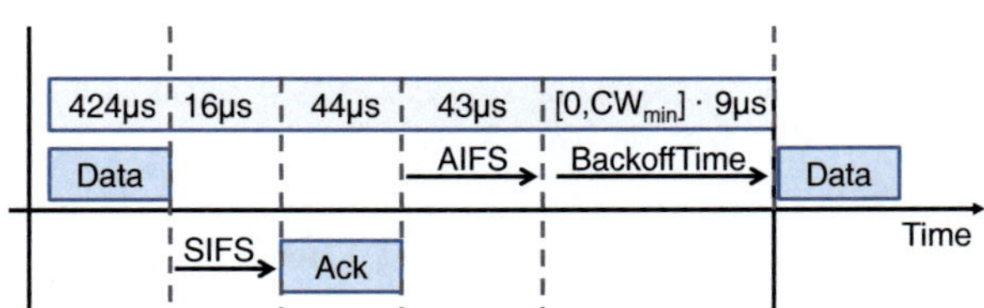

Figure 3.10: One hop duration.

The echo request from MSTA1 is received at MSTA2 after the packet duration t_p where air propagation time is in the order of ns and thus can be neglected[23]. After reception at MSTA2, the echo reply is immediately generated and forwarded to the outgoing queue. However after reception, first an ACK message has to be sent after a Short Interframe Space (SIFS) and the echo reply will be sent after an additional Arbitration InterFrame Space (AIFS). When operating in the 5 GHz band aSIFSTime equals $16\,\mu s$. AIFS is dependent on the used access category (AC) and can be calculated by AIFS = SIFS + AIFSN[AC] · aSlotTime with aSlotTime = $9\,\mu s$ according to [35, Tab. 17-21]. With the default access category best effort (AC_BE) and AIFSn[AC_BE] = 3, AIFS equals $43\,\mu s$. As mentioned before backoff is not performed when

---

[23]As an example we calculate the air propagation time for a distance of $d = 60\,\mathrm{m}$ between the two stations: distance d / speed of light $c = 60\,\mathrm{m}$ / $3 \cdot 10^8\,\mathrm{m/s} = 200\,\mathrm{ns}$.

not using the proposed patch [92] and the size of the contention window (CW) is therefore zero. The total duration thus equals to $2t_p + \text{SIFS} + t_{ack} + \text{AIFS} = 0.951\,\text{ms}$. The total duration of the configurations with 3, 4 and 5 MSTAs can be calculated similarly while each hop adds $2t_p + 2\text{SIFS} + 2t_{ack} + 2\text{AIFS} = 1.054\,\text{ms}$ when not taking backoff into account. As can be seen in Fig. 3.9 this simplified calculation which does not include any backoff does not exactly match the simulation results. In the actual ns-3 implementation for the configuration with 3 to 5 MSTAs backoff is performed only once[24], leading to 16 equally spaced slots separated by the slot time of $9\,\mu s$. This is a result of the non-standard conform simulator implementation. In fact backoff should be calculated at every forwarding attempt except for the first MSTA.

When using the patch proposed by Winlab [92] initially developed for ns-3.16 with the change of the DCF-manager, the simulation shows the expected results (see Fig. 3.11). Backoff is now calculated at every forwarding attempt except for the first MSTA. This patch has been recently included since ns-3.26 which was released on 3rd October 2016.

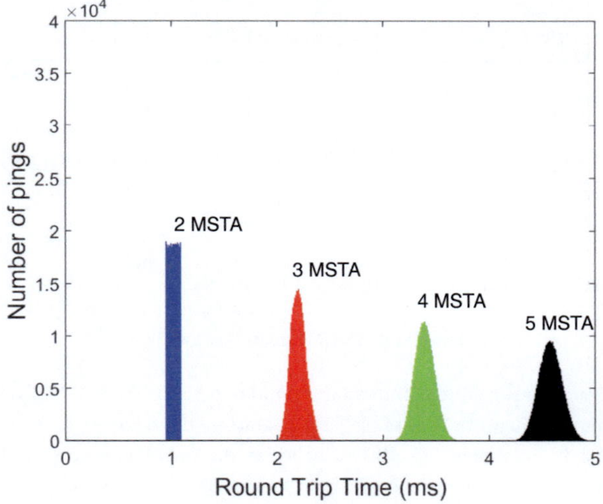

Figure 3.11: Histogram of round-trip-times for the network simulation setups according to Fig. 3.8 with chains of 2, 3, 4 and 5 MSTAs and using the change in DCF-manager for correctly taking into account the required backoff.

The results shown so far lead to the minimum delay per hop caused by the MAC protocol. However, in real environments packet loss is likely to happen due to interference and propagation

---

[24]The backoff procedure is carried out at the first intermediate MSTA after the transmission of the echo reply by the rightmost MSTA because of the detection of a previous concurrency of the echo request sent to the rightmost MSTA. This is the result of the wrong MAC implementation and not further discussed here.

in the wireless medium. Lost packets will result in retransmissions and in an increase of delay and jitter. A lost packet thereby is recognized by a missed ACK at the transmitting MSTA. According to IEEE 802.11 [35, Section 10.3.2.9] the MSTA shall wait for an ACKTimeout interval according to Eq. (3-28), with a value of aRxPHYStartDelay[25] of $20\,\mu s$, which starts at the end of the data transmission. For an OFDM PHY with $20\,MHz$ channel the ACKTimeout equals to $45\,\mu s$.

$$\text{ACKTimeout interval} = \text{aSIFSTime} + \text{aSlotTime} + \text{aRxPHYStartDelay}$$

$$= 16\,\mu s + 9\,\mu s + 20\,\mu s = 45\,\mu s$$

(3-28)

The time introduced per retransmission is shown in Fig. 3.12 and is calculated by[26] ACKTimeout+ AIFS + BackoffTime + t_p.

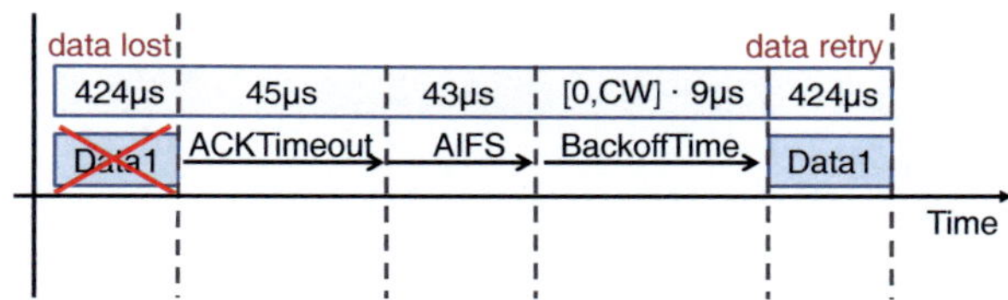

Figure 3.12: Illustration of the additional time interval due to a lost data packet on a single hop.

Thereby BackoffTime is calculated according to Eq. (3-29).

$$\text{BackoffTime} = \text{Random}() \cdot \text{aSlotTime}$$

(3-29)

A random integer is drawn from a uniform distribution over the interval [0,CW], where CW is an integer within the range of values of the PHY characteristics between aCWmin and aCWmax [35, Section 10.3.3]. Assuming AC_BE with the values of aCWmin=15 and aCWmax=1023, the backoff time for the first retransmission is in the interval $[0, 31] \cdot \text{aSlotTime}$. After each retransmission the CW value is increased by sequentially ascending integer powers of 2, minus 1 (see also Appendix A). Therefore the delay of the first retransmission is in the interval $[512, 791]\,\mu s$. An ns-3 simulation has been set up by adjusting the Signal-to-Noise ratio to the effect that the resulting packet success rate (PSR) of the packets with size $b = 300$ bytes (see Tab. 3.3) equals 90%. Figure 3.13 shows the result of the ns-3 simulation in form of a histogram of the round-trip-times for the configuration with 2 MSTAs.

---

[25]It has to be emphasized that this value has been changed in IEEE 802.11-2016 [35, Tab. 17-21] compared to IEEE 802.11-2012 [93, Tab. 18-17] from $25\,\mu s$ to $20\,\mu s$.

[26]This time is smaller than the coherence time expected for the wireless mesh DRS, see also Sec. 3.1.1.2. Thus the assumption of a Rayleigh distribution for the received power within the ACK period of a packet can only be attributed to frequency-selective fading.

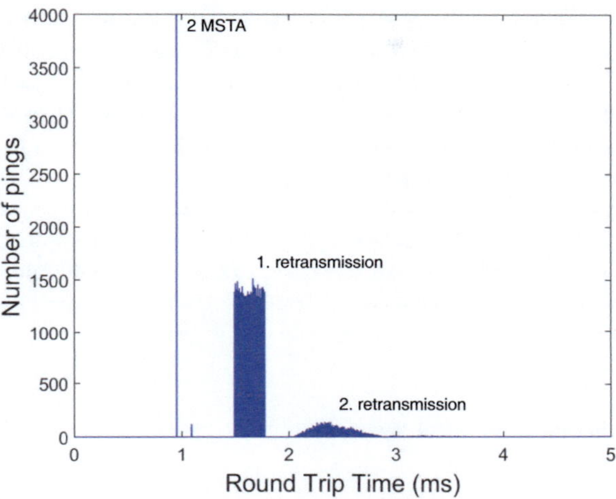

Figure 3.13: Histogram of round-trip-times for the setup according to Fig. 3.8 with 2 MSTAs and a PSR of 90% (change in DCF-manager not considered).

As described previously, at the first transmission no backoff is performed as in this simulation the change in the DCF-manager is not considered. For the first retransmission backoff is calculated leading to 32 equally spaced slots separated by the slot time of $9\,\mu s$. However, the delay with an interval of $[542, 821]\,\mu s$ is not as expected. The difference of about $30\,\mu s$ is a result of an inaccurate implementation in ns-3 which has been also discussed in Bug 761. ns-3 calculates the ACKTimeout interval by Eq. (3-30) which equals to $75.66\,\mu s$ instead of the correct calculation of Eq. (3-28) which explains the difference of about $30\,\mu s$.

$$\text{ACKTimeout\_ns-3} = \text{SIFS} + t_{\text{ack}} + \text{aSlotTime} + 2\,\text{Def\_max\_prop\_delay}$$
$$= 16\,\mu s + 44\,\mu s + 9\,\mu s + 6.66\,\mu s = 75.66\,\mu s$$

(3-30)

A proper patch would have required significant effort and probably be invasive because the current mac-low implementation does not have any knowledge of the time of the RxPHYStart event and therefore remained unresolved. It should be noted that the difference is small and overall simulation results may only lead to a slightly overestimated total delay.

3.2.3 Validation of ns-3 wifi PHY

The ns-3 physical layer model is based on a SINR connection model as introduced in Section 3.1.2.1 and the calculation of Bit Error Rates (BERs) taking into account the forward error correction as defined in IEEE 802.11a. The model calculates the received Signal-to-Interference-plus-Noise-Ratio (SINR) based on parameters used in the simulation model and calculates a packet error rate (PER) based on the mode of operation (e. g. modulation, coding rate) to determine the probability of successfully receiving a IEEE 802.11 frame[27] (denoted as packet success rate (PSR)). The two error rate models which are present in ns-3, the YANS[28] and NIST[29] model are analyzed in the following sections.

Assuming an Additive White Gaussian Noise (AWGN) channel, binary convolutional coding and hard decision Viterbi decoding [94], the PER can be upper bounded by using the equations shown in Sections 3.2.3.2 and 3.2.3.3. Further the calculated PSRs are validated by comparison with the results of an exact simulation of the IEEE 802.11a physical layer in Matlab which is shown in Section 3.2.3.4.

3.2.3.1 Abstraction of physical layer

The implementation of the ns-3 IEEE 802.11a physical layer model follows the implementation described in [88] for both, the YANS and the NIST error rate model. The physical layer is abstracted by using a path-loss propagation model in order to calculate the SINR at the receiver, followed by the calculation of a raw[30] BER covering the used modulation and by deriving a resulting error probability (BER coded) based on the decoding of the convolutional coded bits. From this BER coded the packet error rate PER can be derived, based on the number of bits per packet. Finally, this PER can be realized during simulation of packet reception by drawing a random number from a uniform distribution within the interval $[0, 1]$ and comparing it to the PER (see [88]). If the random number is larger than the PER, then the packet is assumed to be successfully received.

Several path-loss models are available in ns-3, e. g. the Range-Propagation-Loss-Model, the Log-Distance-Path-Loss Model or the COST-Hata-Model (see [95] for a full list of available models), which allows for the calculation of the path loss which is used by the SINR connection model Eq. (3-20). An overview and evaluation of these path loss models is given in [96]. The calculation of the path loss of the ns-3 Log-Distance-Path-Loss Model corresponds to the path loss of the Log-Normal Connection Model in Eq. (3-16). The ns-3 Range-Propagation-Loss-Model will be used in this work in order to allow the comparison of results with system level simulations carried out in Matlab which are using the unit disk connection model. Using this model allows that receivers at or within a maximum range in meters to receive the transmission at

---

[27]Packets or datagrams are normally units of data in the network layer (IP in case of the Internet). As we are focusing on the PHY performance here namely the PER on layer 1 of IEEE 802.11 has to be considered. Units of data on layer 1 are denoted as frames. For analyzing error rates on layer 1 the term packet and hence PER is usually used as synonym for a frame which is also used throughout this work.

[28]Yet Another Network Simulator (YANS), [88].

[29]National Institute of Standards and Technology (NIST), [89].

[30]The capacity of a medium to carry data is described as the raw data bandwidth of the media. The corresponding BER of the raw data packet is the raw BER.

the transmit power level. Beyond the maximum range the received power is set to $-1000\,\text{dBm}$ (effectively zero). Thus nodes beyond the maximum range don't contribute to the interference noise caused by concurrent transmissions when calculating the SINR in Eq. (3-20). Concurrent transmissions within the transmission range will most likely lead to lost packets because of the same transmit power level which is used for all MSTAs. This allows for simulation results of the SINR connection model which correspond to the unit disk connection model introduced in Section 3.1.2.1 with the additional restriction that concurrent packet transmission within the range of a given MSTA are almost impossible. The SINR thereby is calculated in the ns-3 InterferenceHelper class.[31]

Let's have a closer look on the ns-3 wifi PHY implementation. According to the description in [88] the PHY can be in one of four states:

- TX: the PHY is currently transmitting a signal.

- SYNC: the PHY is synchronized on a signal and is waiting until it has received the last bit to forward it to the MAC.

- CCA_BUSY: the PHY is not in the TX or SYNC state, but the energy which is measured on the medium is higher than the Energy Detection Threshold (EDThreshold) (as defined by the Clear Channel Assessment (CCA) requirements [35, Sec. 17.3.10.6]). A frame k can be received if its $\text{SINR}(k,t)$ level is above the threshold for time t.

- IDLE: the PHY is not in the TX, SYNC or CCA_BUSY states, but the behavior is the same as in the CCA_BUSY state. A frame k can be received if its $\text{SINR}(k,t)$ level is above the threshold for time t.

A received frame is dropped if the first bit of this frame is received while the PHY is in the TX or the SYNC state. If the PHY is in the IDLE or CCA_BUSY state, the received energy of the first bit will be calculated and compared to the EDThreshold value. If the energy of the first bit of the frame k is higher than the EDThreshold value, the PHY moves to the SYNC state and schedules an event when the last bit of this frame is expected to be received. Otherwise, the PHY either switches to CCA_BUSY if the total energy of the received frame reached EDTreshold or stays in IDLE state. In both cases, the frame is dropped [88].

The power of the received signal $S(k,t)$ is assumed to be zero outside of the reception interval of the frame k whereas inside the reception interval of the frame k it is calculated from the transmission power with a path-loss propagation model as described in Section 3.1.2.1. When the last bit of a frame upon which the PHY is synchronized is received, the probability that the frame is received with any error, $P_{\text{err}}(k)$, is calculated in order to decide whether or not this frame could be successfully received or not. Then a random number rnd is drawn from a uniform distribution in the interval [0,1] and is compared against $P_{\text{err}}(k)$. If rnd is larger

---

[31]In Chapter 8 the SINR model will be used with a new ns-3 model developed in this work which accounts for the path loss of the ITU Urban Micro cell (UMi) NLOS path loss model.

than $P_{\text{err}}(k)$, the frame is assumed to be successfully received. Otherwise, it is reported as an erroneous reception [88].

The power of the received signal $S(k,t)$ can be calculated as shown in Section 3.1.2.1 by Eq. (3-17). This equation is reconsidered here and rewritten for the reception of a frame k at time t. The interference noise of the subset of nodes T_i that are simultaneously transmitting with node i as shown in Eq. (3-20) will be denoted as $N_i(k,t)$ which represents the interference noise, that is, the sum of the power of all the other signals received on the same channel. The Signal-to-Interference-plus-Noise-Ratio SINR(k,t) for frame k is calculated[32] according to [88]:

$$\text{SINR}(k,t) = \frac{S(k,t)}{N_i(k,t) + N_f} \tag{3-31}$$

where N_f represents the receiver noise floor[33]. It should be noted that the PHY preamble, PHY header and payload of the frame are separated for the calculation of the SINR, since the payload may be encoded with a higher order modulation. The SIGNAL field of the PHY header is always encoded with the most robust modulation BPSK and a half-rate convolutional code which leads to different BERs. Additionally, the SINR and the resulting BERs have to be calculated for several chunks because of the change of the interference noise during the reception of a frame. This can be explained by Fig. 3.14: The signal power of a concurrent transmission of frame b is interpreted as interference noise for the reception of frame a. Because of the change of the interference noise during the reception of frame a the calculation of the successful reception of frame a will be divided into several chunks of size l_i.

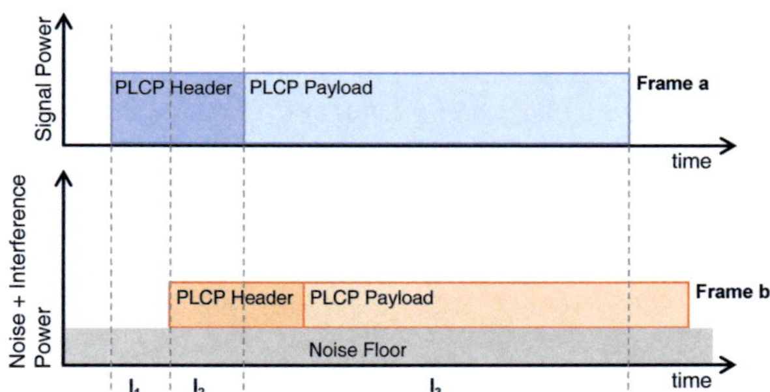

Figure 3.14: Sketch of the successful reception of frame a which is divided into several chunks of size l_i for the calculation of the SINR and the resulting PER. [88] [97]

---

[32] This equation is implemented in the ns-3 InterferenceHelper class and used for all simulations throughout this work. The received signal power $S(k,t)$ and interference noise power $N_i(k,t)$ will be calculated by using an appropriate path loss model, e. g. the Range-Propagation-Loss-Model, the Log-Distance-Path-Loss Model or the ITU Urban Micro cell (UMi) NLOS path loss model.

[33] $N_f = k_B\, T\, B\, F$ with Boltzman constant k_B, temperature T, channel bandwidth B and noise factor of the receiver F.

Additionally, the preamble and payload of the frame are separated into different chunks as already stated. The calculation of the overall packet error rate $P_{\text{err}}(k)$ for all chunks of bits which are located at the intervals l_i of the frame k will be detailed in the following section. It should be noted that if frame b has a stronger signal than frame a the PHY is unable to tune to the stronger frame in the current ns-3 implementation. The reason is the so called frame-capture effect which is currently not implemented in the PHY model. The frame capture effect is a feature which allows devices to switch to a stronger signal during the reception of a weaker frame. Due to the additional interference and change of the SINR, the currently received packet reception of frame a will most likely fail if frame b has a stronger signal than frame a depending on the region of overlap and power levels. The stronger frame is lost as well. [97] shows an evaluation of the ns-3 interference model. In particular the capability to successfully decode OFDM-frames which are overlapped in time by one or more lower-power frames received on the same carrier frequency has been analyzed by measurements and simulations in ns-3. The Frame Error Ratio (FER) is observed in a simple two-frame-overlap simulation experiment for several Signal-to-Interference Ratios (SIR) and further compared with measurement results. The measured frame error ratio shows the same behavior as the ns-3 simulation when the higher power frame (frame a) starts before the lower power frame (frame b) (compare Fig. 3.14), however a systematic difference of 4 dB in the required SIR is seen. One possible explanation which is detailed in [97] is that the OFDM signal cannot be treated exactly like Additive White Gaussian Noise (AWGN) noise which leads to a pessimistic overestimation of the OFDM interference because the energy within the OFDM signal might not be uniformly distributed over the channel bandwidth. However, it should also be noted that the used NIST error rate model in the carried out ns-3 simulations in [97] shows overly pessimistic results of about 2 dB which is shown in Section 3.2.3.4 of this work and would also explain part of the systematic error shown in [97].

The capture effect which is not implemented in ns-3 could be observed by the measurements when the higher power frame (frame a) starts after the lower power frame (frame b). Interestingly a capture probability plateau was seen at high SIR values whereby the percentage of overlap of the two frames seemed not to matter anymore. It is stated in [97] that this might be caused by timing estimation errors at the moment of resynchronization which dominate over interference power contributions during the frame overlap phase. See [97] for further results and explanation.

A not considered capture effect may lead to overly pessimistic results in high load scenarios. For the low load scenarios considered in this work the impact of the capture effect is assumed to be very low.

The two error rate models available in ns-3 which cover the calculation of the corresponding bit error and packet error rates for a chunk of bits located in interval l_i are discussed in more detail in the following sections.

3.2.3.2 YANS error rate model

According to the YANS model [88], the BER p_{BPSK} for Binary Phase-Shift Keying (BPSK) and p_{QAM} for Quadrature Amplitude Modulation (QAM) for a packet k at time t are calculated by Eqns. (3-32) to (3-35). t thereby denotes the discrete time of a bit. The bit error rate for the M-QAM-system[34] is calculated by transferring the M-QAM system to an equivalent $\sqrt{M}$-PAM-system (Pulse Amplitude Modulation). The calculation according to Eq. (3-33) equals the actual implementation in ns-3, however, is corrected here with respect to what is presented in [88] by a factor $1/\log_2(M)$ in the error probability for a single symbol in order to correctly take into account the number of bits per QAM symbol. $P_{\sqrt{M}}$ is the probability of error of a $\sqrt{M}$-ary PAM system with half of the average power found in each quadrature signal of the equivalent QAM system [98]. Following common notation E_b denotes the available energy per bit where N_0 represents the noise power spectral density and $\frac{E_b}{N_0}$ is substituted by Eq. (3-36) in the ns-3 implementation.

$$p_{\text{BPSK}}(k,t) = \frac{1}{2}\text{erfc}\left(\sqrt{\frac{E_b(k,t)}{N_0}}\right) \tag{3-32}$$

$$p_{\text{QAM}}(k,t) = \frac{1 - (1 - P_{\sqrt{M}}(k,t))^2}{\log_2(M)} \tag{3-33}$$

with

$$P_{\sqrt{M}}(k,t) = \left(1 - \frac{1}{\sqrt{M}}\right) X(k,t) \tag{3-34}$$

and

$$X(k,t) = \text{erfc}\left(\sqrt{\frac{1.5}{M-1}\log_2(M)\frac{E_b(k,t)}{N_0}}\right) \tag{3-35}$$

The calculation of the SINR per bit in the YANS model is shown in Eq. (3-36). $\frac{E_b}{N_0}$ is calculated by the fraction of the channel bandwidth B_t (e.g. 20 MHz) over the raw bitrate[35] R_b of the transmission mode used by packet k at time t multiplied by the Signal-to-Interference-plus-Noise-Ratio (SINR) of packet k at time t.

$$\frac{E_b(k,t)}{N_0} = \text{SINR}(k,t)\frac{B_t}{R_b(k,t)} \tag{3-36}$$

R_b is calculated by the number of bits per OFDM symbol over the symbol interval time. For example this leads to the following result for BPSK[36], Eq. (3-37).

---

[34]Note that IEEE 802.11 is using Gray-coded constellation mappings [35, Chapter 17].

[35]Raw bits denote the bits transmitted on the PHY as described at the beginning of Section 3.2.3.1.

[36]For an IEEE 802.11a OFDM system [35, Tab. 17-5] we have: the number of data subcarriers $N_{\text{data}} = 48$, number of pilot subcarriers $N_{\text{pilot}} = 4$ and duration of the SIGNAL BPSK-OFDM symbol time $4\mu s$.

$$R_b = \frac{\text{Bits / OFDM Symbol}}{\text{Symbol Interval Time}} = \frac{48}{4\mu s} = 12\text{Mbit/s} \tag{3-37}$$

This formulation doesn't take into account the reduction of energy due to the cycling prefix and the reduction of the net energy due to the pilot carriers which do not transport information which will be further discussed in Sec. 3.2.3.4.

For each interval l_i where $p(k,t) = p_i$ (p_i denotes the bit error probability p_{BPSK} for BPSK or p_{QAM} if QAM is used) is constant, which means that $R_b(k,t)$ and $\text{SINR}(k,t)$ are constant within this interval, $P_e(k,l_i)$ can be calculated which represents an upper bound on the probability that an error is present in the chunk of bits located in interval l_i for the packet k. Assuming an Additive White Gaussian Noise channel, binary convolutional coding and hard-decision Viterbi decoding, $P_e(k,l_i)$ can be calculated by the Eqns. (3-38) to (3-41) as detailed in [98]. Lets briefly describe this method of calculating $P_e(k,l_i)$.

Since convolutional codes are linear codes, the probability of error can be obtained by assuming that the all-zero sequence is transmitted and then determining the probability that the decoder decides in favor of a different sequence. To this end we first consider the probability that the decoder decides for an erroneous path within the trellis instead of the all-zero path, where the erroneous path differs in d raw bits from the all-zero path and thus has a Hamming distance of d from the all-zero path. We call this probability $P_2^u(d,k,l_i)$ the pairwise error probability. The pairwise error probability $P_2^u(d,k,l_i)$ of incorrectly selecting a path within the trellis when the Hamming distance d between the correct and the incorrect path is even or odd is calculated by Eq. (3-38). [98]

$$P_2^u(d,k,l_i) = \begin{cases} \displaystyle\sum_{j=\frac{d+1}{2}}^{d} \binom{d}{j} p_i^{\,j}(1-p_i)^{d-j} & (d \text{ odd}) \\[3em] \displaystyle\sum_{j=\frac{d}{2}+1}^{d} \binom{d}{j} p_i^{\,j}(1-p_i)^{d-j} + \frac{1}{2}\binom{d}{\frac{d}{2}} p_i^{\frac{d}{2}}(1-p_i)^{\frac{d}{2}} & (d \text{ even}) \end{cases} \tag{3-38}$$

This can be easily understood by noting, that a decision for an incorrect path with Hamming distance d to the all zero path could only occur if – for d odd – at least $\frac{(d+1)}{2}$ errors have been introduced by the channel. The same holds for d even with an additional probability of $\frac{1}{2}$ for choosing the incorrect path if the channel has introduced $\frac{d}{2}$ errors.

The overall performance achieved by the Viterbi decoding algorithm in the receiver then is estimated by calculating the first-event-error probability $P_u(k,l_i)$. There is no simple exact expression for the first-event error probability $P_u(k,l_i)$, but the error probability can be over-bounded by the sum of the pairwise error probabilities $P_2^u(d,k,l_i)$ of the chunk of bits located in interval l_i for the packet k over all possible paths that merge with the all-zero path at the given state within the trellis diagram [98]. As there are for each Hamming distance d greater or equal to the free distance d_{free} of the code many possible paths that merge with the all-zero path at a given state, each pairwise error probability $P_2^u(d,k,l_i)$ has to be weighted by the number of paths α_d having equal Hamming distance d. With this calculation scheme the so-called union bound can be obtained [98] which is used in the YANS model [88]. The coefficients α_d represent for each d the number of paths with distance d that merge with the all-zero path. According to the IEEE 802.11 OFDM PHY specification [35, Sec. 17.3.5.6] the convolutional encoder shall use the industry-standard generator polynomials, $g_0 = 133_8$ and $g_1 = 171_8$, of rate $R = 1/2$. With $d = d_{\text{free}} = 10$ the parameter $\alpha_{d=d_{\text{free}}=10} = 11$ indicates that there are 11 paths of Hamming distance $d = 10$ from the all-zero path that merge with the all-zero path at a given node of the trellis diagram. The coefficients α_d can be obtained from [99]. Using these considerations we can calculate the so called union bound:

$$P_u(k, l_i) < \sum_{d=d_{\text{free}}}^{\infty} \alpha_d P_2^u(d, k, l_i) \tag{3-39}$$

The union bound (Eq. (3-39)) is calculated in the YANS model by the sum over a finite number of coefficients of α_d given by the number of Hamming distances[37] K.

$$P_u(k, l_i) < \sum_{d=d_{\text{free}}}^{d_{\text{max}}} \alpha_d P_2^u(d, k, l_i) \tag{3-40}$$

The union bound (Eq. (3-40)) is calculated for BPSK with $d_{\text{max}} = d_{\text{free}}$ (i. e. $K = 1$) and for QAM with $d_{\text{max}} = d_{\text{free}} + 1$ (i. e. $K = 2$).

Finally, the upper bound on the probability $P_e(k, l_i)$ that an error is present within the chunk of bits located in interval l_i for the packet k is calculated by using Eq. (3-41) with the number of raw bits L used in the ns-3 implementation. Let us repeat that raw bits denote the bits transmitted on the PHY and thus the number of raw bits can be calculated by the number of data bits divided by the code rate ($R = 1/2$, $2/3$ or $3/4$ [35, Tab. 17-16]).

$$P_e(k, l_i) \leq 1 - (1 - P_u(k, l_i))^{L(k, l_i)} \tag{3-41}$$

The packet error rate $P_{\text{err}}(k)$ for a packet k is finally calculated by the product of the calculated upper bounds on the probability that an error is present $P_e(k, l_i)$ within the chunk of bits for all intervals l_i of the packet k. This is expressed by Eq. (3-42).

$$P_{\text{err}}(k) = 1 - \prod_i (1 - P_e(k, l_i)) \tag{3-42}$$

3.2.3.3 NIST error rate model

For the calculation of the BER in the ns-3 NIST model [89] the Eqns. (3-44) to (3-47) are used. These equations represent an approximation of Eqns. (3-33) to (3-35) with inserted values of the number of constellation points M. Eq. (3-44) is corresponding to Eq. (3-32) of

---

[37] K is the number of coefficients α_d used for the calculation of the error bound.

the YANS model. The SINR in Eq. (3-43) represents the Signal-to-Noise-Ratio (SNR) per symbol and doesn't account for the ratio of used subcarriers in the OFDM system as used in the YANS model in Eq. (3-36) by the fraction of the channel bandwidth B_t over the raw bitrate R_b.

$$\frac{E_b(k,t)}{N_0} = \frac{\text{SINR}(k,t)}{\log_2(\text{M})} \tag{3-43}$$

$$p_{\text{BPSK}}(k,t) = \frac{1}{2}\text{erfc}\left(\sqrt{\text{SINR}(k,t)}\right) \tag{3-44}$$

$$p_{\text{QPSK}}(k,t) = \frac{1}{2}\text{erfc}\left(\sqrt{\frac{\text{SINR}(k,t)}{2}}\right) \tag{3-45}$$

$$p_{16\text{-QAM}}(k,t) = \frac{3}{4\cdot2}\text{erfc}\left(\sqrt{\frac{\text{SINR}(k,t)}{5\cdot2}}\right) \tag{3-46}$$

$$p_{64\text{-QAM}}(k,t) = \frac{7}{8\cdot3}\text{erfc}\left(\sqrt{\frac{\text{SINR}(k,t)}{21\cdot2}}\right) \tag{3-47}$$

In the ns-3 NIST model Eq. (3-48) and (3-49) are used for the calculation of an upper bound $P_b^c(k,l_i)$ on the probability of a bit error instead of the first-event error probability $P_u(k,l_i)$ as used in the YANS model. Using the upper bound $P_2^c(d,k,l_i)$ which is also called the Chernoff Bound in Eq. (3-48) instead of the expressions in Eq. (3-38) yields a looser upper bound $P_b^c(k,l_i)$ on the probability of a bit error (see also [98, Section 8.2.4]). Additionally, the NIST model is using the multiplication factors β_d which correspond to the number of nonzero information bits that are in error when an incorrect path is selected for the specified Hamming distance d. This is in contrast to the YANS model which uses the multiplication factors α_d denoting the number of paths with given Hamming distance d that merge with the all-zero path. The difference is that in each of the paths denoted by α_d several information bits can be in error which are given by β_d. p_i again is the bit error probability for the different modulations where p_i are constant for each chunk l_i. The coefficients β_d can be obtained from[38] [99].

$$P_2^c(d,k,l_i) < [4p_i(1-p_i)]^{d/2} \tag{3-48}$$

$$P_b^c(k,l_i) < \frac{1}{2b}\sum_{d=d_{\text{free}}}^{\infty}\beta_d P_2^c(d,k,l_i) \tag{3-49}$$

Another difference between the ns-3 implementation of the two models lies in the fact, that the YANS model is using the union bound with $K = 1$ for BPSK and $K = 2$ for QAM and the NIST model is using the Chernoff bound with $K = 10$, where $K = (d_{\text{max}} - d_{\text{free}}) + 1$ again defines the maximum Hamming distance d_{max} considered in the sum and thus defines the number of coefficients β_d or equivalently the number of nonzero information bits that are in error when an incorrect path is selected. Beside that, the SINR is calculated in a different manner which leads to an additional shift of the resulting packet success rates (PSRs).

---

[38] $b = (1, 2, 3)$ for code rates $(1/2, 2/3, 3/4)$. The additional factor of $1/2$ which is presented in the NIST model [89], but not in [98] or [94] is another inaccuracy which unfortunately could not be followed up.

The packet error rate $P_e^c(k, l_i)$ for the chunk of bits found in the interval l_i of the packet k is then calculated as shown in the YANS model by using Eq. (3-50) with the number of raw bits $L(k, l_i)$.

$$P_e^c(k, l_i) \leq 1 - (1 - P_b^c(k, l_i))^{L(k, l_i)} \tag{3-50}$$

The packet error rate $P_{\text{err}}(k)$ for a packet k finally is calculated again as shown in the YANS model by the product of the upper bound $P_e^c(k, l_i)$ calculated by Eq. (3-50) for all chunks of bits l_i (see Eq. (3-51)).

$$P_{\text{err}}(k) = 1 - \prod_i (1 - P_e^c(k, l_i)) \tag{3-51}$$

3.2.3.4 Validation of the error rate models

The error-rate models implemented in ns-3 are validated by using a Matlab implementation of the equations shown in Sections 3.2.3.2 and 3.2.3.3 and by comparing the results to an exact physical-layer simulation of an IEEE 802.11a OFDM transceiver system operating on complex time samples which was implemented in Matlab. The block diagram[39] of this implementation is shown in Fig. 3.15.

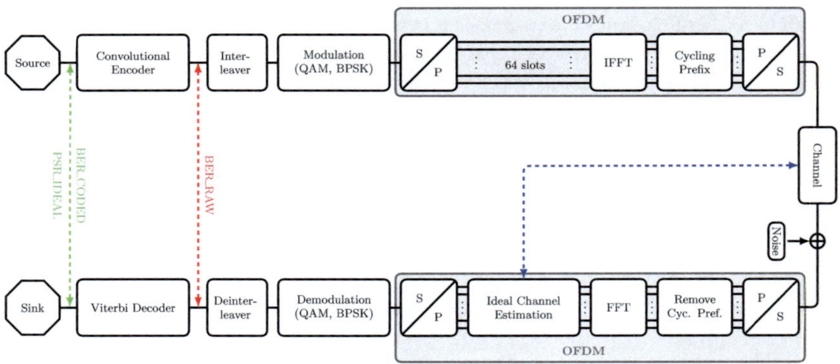

Figure 3.15: Exact physical layer simulation in Matlab of an IEEE 802.11a OFDM communication link.

It has to be noted that the ns-3 implementation of the YANS model which is using the union bound (Eqns. (3-38) and (3-39)) and of the NIST model which is using the Chernoff bound (Eqns. (3-48) and (3-49)), detailed in Sections 3.2.3.2 and 3.2.3.3, are using an inaccurate calculation of the corresponding SNR and a wrong calculation of the PSR by an inadequate

---

[39] A scrambler is not used because the data source is modeled as a PRBS generator.

number of bits L (Eqns. (3-41) and (3-50)). In this section a revised version of the error rate model is proposed, using the upper bound (Eq. (3-56)), a correct representation of the SNR (Eq. (3-55)) and the correct number of data bits for the calculation of the PSR equivalently to Eq. (3-50) instead.

The relation between the available energy per bit E_b and the transmitted energy per OFDM symbol E_{OFDM} is described in [100]. E_{OFDM} is the energy distributed along the whole symbol, including signalling overheads. E_b thus can be expressed as:

$$E_{\mathrm{b}} = E_{\mathrm{OFDM}} \left(\frac{N_{\mathrm{FFT}}}{N_{\mathrm{FFT}}+N_{\mathrm{CP}}} \right) \left(\frac{N_{\mathrm{data}}}{N_{\mathrm{data}}+N_{\mathrm{pilot}}} \right) \left(\frac{1}{N_{\mathrm{data}}N_{\mathrm{BPSCS}}R} \right) \tag{3-52}$$

N_{FFT} is the FFT length[40]. N_{CP} is the length of the cycling prefix (CP) expressed as the number of time samples in the cyclic prefix at an oversampling rate of 1. N_{data} is the number of data subcarriers and N_{pilot} the number of pilot subcarriers respectively. N_{BPSCS} is the number of raw bits per symbol in each OFDM subcarrier and R the code rate of the Forward Error Correction (FEC). Based on the relation in Eq. (3-52) the Signal-to-Noise-Ratio (SNR) and $\frac{E_b}{N_0}$ (SNR per bit or the ratio of energy per bit to the one-side noise spectral density N_0) can be determined. The SNR is defined as the ratio of signal power $\mathcal{P}_{\mathrm{signal}}$ to noise power where the signal power is the energy (variance) per time sample.[41] [100]

$$\mathcal{P}_{\mathrm{signal}} = \frac{E_{\mathrm{OFDM}}}{(N_{\mathrm{FFT}}+N_{\mathrm{CP}})\, t_s} = \frac{E_{\mathrm{OFDM}}\, B_t}{N_{\mathrm{FFT}} + N_{\mathrm{CP}}} \tag{3-53}$$

Therefore using Eqns. (3-52) and (3-53) the SNR can be calculated as:

$$\mathrm{SNR} = \frac{\mathcal{P}_{\mathrm{signal}}}{N_0\, B_t} = \frac{E_{\mathrm{OFDM}}}{N_{\mathrm{FFT}} + N_{\mathrm{CP}}} \cdot \frac{1}{N_0} = \frac{E_{\mathrm{b}}}{N_0} \left(\frac{N_{\mathrm{data}}+N_{\mathrm{pilot}}}{N_{\mathrm{FFT}}} \right) N_{\mathrm{BPSCS}}\, R \tag{3-54}$$

The representation in the YANS model, Eq. (3-36), doesn't take into account the energy per bit to noise power spectral density ratio $\frac{E_b}{N_0}$ correctly which is used for the BER calculations. $\frac{E_b}{N_0}$ is calculated in Eq. (3-36) by the raw bit rate R_b (calculated by the number of bits per OFDM symbol) over the symbol interval time. This formulation neither takes into account the reduction of energy due to the cycling prefix (first term in Eq. (3-52)) nor the reduction of the net energy due to the pilot carriers which do not transport information (second term in Eq. (3-52)) [100]. On the other hand Eq. (3-43) of the NIST model doesn't account for the ratio of used subcarriers in the OFDM system and CP at all.

Eqns. (3-32) to (3-35) can be used in order to calculate the bit error probability p_{BPSK} for BPSK and p_{QAM} for QAM, where, however, in order to correctly calculate the raw bit error probability, the raw $\frac{E_b}{N_0}$ (Eq. (3-55)) has to be used which is the equivalent to Eq. (3-54) divided by code rate[42] R.

---

[40]For an IEEE 802.11a OFDM system $N_{\mathrm{FFT}} = 64$ (FFT – Fast Fourier Transform), $N_{CP} = 16$ with a guard interval of $0.8\mu s$, $N_{\mathrm{data}} = 48$, $N_{\mathrm{pilot}} = 4$, $N_{\mathrm{BPSCS}} = 1,2,4,6$ for BPSK, QPSK, 16-QAM and 64-QAM respectively, $R=1/2,\ 2/3,\ 3/4$.

[41]As stated earlier B_t is the channel bandwidth and t_s the sampling time.

[42]It has to be noted that interference is assumed to be zero in the equations and the term SNR is used in the calculations instead of SINR.

$$\left(\frac{E_\mathrm{b}}{N_0}\right)_{\mathrm{raw}} = \mathrm{SNR}\left(\frac{N_\mathrm{FFT}}{N_\mathrm{data}+N_\mathrm{pilot}}\right)\left(\frac{1}{N_\mathrm{BPSCS}}\right) \tag{3-55}$$

The performance achieved by the Viterbi decoding algorithm in the receiver then can be estimated by calculating the upper bound $P_b^u(k,l_i)$ on the bit error probability. The probability of incorrectly selecting a path when the Hamming distance d is even or odd is shown in the YANS model (see Eq. (3-38)). As discussed above the error probability can be overbounded by the sum of the pairwise error probabilities $P_2^u(d,k,l_i)$ over all possible paths that merge with the all-zero path at the given node. The multiplication factors α_d used for the calculation of the union bound in the YANS model correspond to the number of paths that merge with the all-zero path for each d out of the set of distances d. Instead the multiplication factors β_d which corresponds to the number of nonzero information bits that are in error when an incorrect path is selected for the specified Hamming distance d to the all-zero path have to be used to obtain the so called upper bound on the probability of a bit error. The coefficients β_d and d_free for punctured Codes can be taken from [101]. We thus obtain the upper bound for the error probability according to [94]:

$$P_b^u(k,l_i) < \frac{1}{b} \sum_{d=d_\mathrm{free}}^{\infty} \beta_d P_2^u(d,k,l_i) \tag{3-56}$$

The packet error rate $P_\mathrm{err}(k)$ for a packet k is then calculated as shown in the YANS implementation by Eq. (3-41) and Eq. (3-42) or as shown in the NIST implementation by Eq. (3-50) and Eq. (3-51) where the number of data bits L has to be used instead of the number of raw bits.

Figure 3.16 shows the results of the ns-3 implementations for the YANS and NIST model and the calculation of the union bound, upper bound and Chernoff bound for $K=1$ and $K=10$ for BPSK and $R=1/2$. The ns-3 simulations have been carried out by sending 81.000 packets. The SNR in the Matlab implementation is calculated by using Eq. (3-55). The bit error probability p_{BPSK} is calculated as in the YANS and NIST models shown in Eqns. (3-32) and (3-44). The union bound is calculated by Eqns. (3-38) and (3-39), the upper bound is calculated by Eqns. (3-38) and (3-56) and the Chernoff bound is calculated by Eqns. (3-48) and (3-56) respectively. By using Eq. (3-56) also for the calculation of the Chernoff bound the inaccurate factor of $1/2$ which is presented in the NIST model is neglected. The packet success rate (PSR) for the union, upper and Chernoff bound is then calculated by using Eq. (3-50) with the number of data bits $L=8000$ bits while the ns-3 YANS and NIST implementation are incorrectly calculated with the number of raw bits.

The upper and Chernoff bound vary less than $1\,\mathrm{dB}$ for BPSK with code rate $R=1/2$. The ns-3 YANS model is using the union bound with $K=1$ and the NIST model is using the Chernoff bound with $K=10$. Because of the inaccurate SNR calculation and the wrong number of bits for the calculation of the PSR (raw bits instead of data bits which are twice as many for $R=1/2$) the YANS model is overly optimistic (about $2\,\mathrm{dB}$) whereas the NIST model is

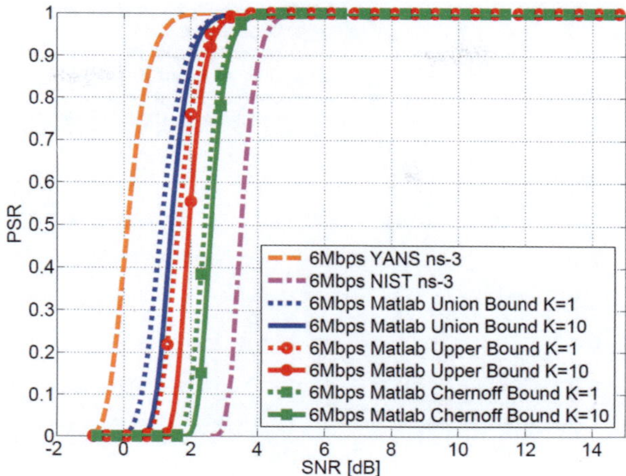

Figure 3.16: Comparison of ns-3 and Matlab implementations for calculating the packet success rate (PSR).

pessimistic (about 2 dB) when compared to the Matlab implementations using the respective bounds used in the YANS and NIST implementations.

The results outlined above indicate the following changes to the ns-3 implementation of the YANS and the NIST model: First the number of raw bits used in Eq. (3-50) must be multiplied by the code rate R in order to get the number of data bits. Second the calculation of the SNR must be changed to the representation in Eq. (3-55). Third it is recommended to use the upper bound by using the multiplication factors β_d for the calculation instead of the Chernoff bound or the union bound.

The exact simulation of the physical layer for an OFDM link according to IEEE 802.11a includes convolutional coding, interleaver, modulation and cycling prefix at the transmitter as well as ideal channel estimation, demodulation, deinterleaver and hard-decision Viterbi decoder at the receiver (Fig. 3.15). The traceback length of the Viterbi decoder is 35 in all simulations which is 5 times the constraint length of the convolutional code with the generator polynomial [133,171]. The SNR in this exact simulation of the physical layer is a direct input parameter and is realized as additive white Gaussian noise on the OFDM time signal. This SNR is related to the $\frac{E_b}{N_0}$ of the data bits by Eq. (3-54). Figure 3.17 shows the result of the exact physical layer simulation in Matlab using an AWGN channel. The simulation has been carried out using 10^7 symbols. The result asymptotically is approached by the upper bound, which is illustrated by $K = 10$ and thus confirms the calculation.

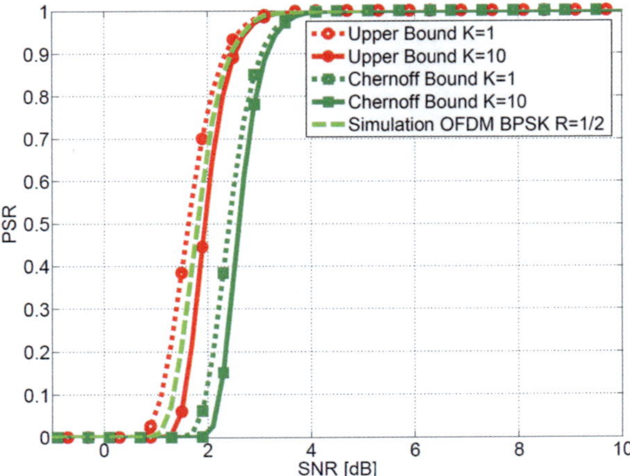

Figure 3.17: Comparison of the BPSK PSR from Upper Bounds and exact Matlab simulation of the physical layer for an IEEE 802.11a communication system.

3.3 Throughput and Delay Performance of Wireless Mesh Networks

Most of the current research results addressing VoIP in WMN (e. g. [102], [103] and [104]), are focusing on the capacity of the mesh network in terms of a maximum number of calls while still having sufficient QoS. In the use case of a disaster recovery system, however, we are more interested in low-rate data services and a low number of VoIP calls in the WMN providing still sufficient QoS over a maximum number of hops.

The QoS requirements of VoIP according to [105] thereby are:

- Packet loss should be no more than 1 percent.

- One-way latency (mouth to ear) should be no more than 150 ms.

- Average one-way jitter should be targeted at less than 30 ms.

Fulfilment of the requirements on packet loss, latency and jitter are dependent on the maximum number of hops carried out in the mesh network. For the analysis of the latency also processing delays of real hardware have to be taken into account. The analysis presented in the existing work is based on testbed measurements, analytical models as well as simulation. [102] focuses on call admission control and different routing strategies based on available capacity or call statistics. The capacity utilization model and call admission decisions presented in [102] are evaluated on an experimental testbed based on six identical Dell laptops running Linux 2.6.15, and using Atheros 802.11a/b/g cards with madwifi driver while routing was evaluated in ns-2. [103] showed that the number of calls supported in a 15 node mesh setup – while each node has a 802.11b 2 Mbps link – is reduced from 8 calls in a single hop scenario to one call after 5 hops in their testbed. By providing a multihop aggregation mechanism, which has been also validated by ns-2 simulations, the available capacity could be increased. [104] determined the maximum number of supported calls in an experimental WMN testbed, based on Mesh Routers running DD-WRT Linux software, at different hop counts. After 5 hops only one call was possible with sufficient performance by using 11Mbps links between the Mesh Routers. Let us emphasize that none of the previously mentioned papers used the amendment for wireless mesh networking IEEE 802.11s in their testbeds, but IP-layer based routing algorithms. An estimation of the processing delay and determination of a maximum number of hops was also not part of their research.

Some papers (e. g. [106], [107] and [108]) propose analytical models based on Markov chains, initially proposed by Bianchi [109], to compute the throughput performance of the IEEE 802.11 Distributed Coordination Function (DCF). Those papers focus mainly on saturated throughput which is not applicable in the use case of this thesis. [106] analyzed the saturation performance of the IEEE 802.11e Enhanced Distributed Channel Access (EDCA) throughput and in addition also the performance of delay and delay jitter. [107] presented analytical formulas also for the throughput and access delay in non-saturated mode, which are obtained and verified by ns-2 simulations. Recently, [108] analyzed the end-to-end throughput and delay for a IEEE 802.11 string topology multihop network using a Markov chain model. However, channel conditions of all the links are ideal and transmission failures occur only due to frame collisions in their model. All previously mentioned papers account for delay in terms of the access delay which means the time from when a data frame at the transmit station starts contending for the channel to the time the data reaches the MAC layer of the receiver. The processing delay introduced by real hardware is not included in these models.

[110], [111] and [112] analyze the performance of VoIP in WMN by simulative approaches. [110] presented the simulation results of a VoIP WMN based on simulations with QualNet. In their simulations average one way delay increased linearly up to 7 mesh nodes. However, after 7 mesh nodes the delay increases abruptly which is not further explained in this paper. In [111] the performance of VoIP by using three different routing protocols is presented in the context of IEEE 802.11s where the HWMP protocol was implemented in ns-2. [112] evaluates the Enhanced Distributed Channel Access mechanism by simulation of topologies which considered different types of delay sensitive applications such as VoIP. The results presented in [112] showed that QoS is not guaranteed in an IEEE 802.11s WMN since EDCA fails to categorize different types of applications into suitable QoS classes.

It should be noted, that the presented simulation-based research (e. g. [110], [111] and [112]) doesn't account for hardware induced processing delays resulting from software execution time and kernel scheduling. Recently, [113] proposed a methodology to extract service models from real multi-threaded Operating Systems (OSs) by means of instrumentation and tracing. A proof of concept implementation of service models for IP forwarding in ns-3, derived from a Google Nexus One and Nokia N900 smart phone, showed the possibility to account for the overhead of protocol processing. [114] developed a simulation model in ns-3 for packet processing via the Linux network stack. They have analyzed the Network Interface Control (NIC) driver and NAPI (New API) as part of the OS and have modeled and simulated the NIC driver and NAPI mechanism by extending their ns-3 resource management module. Currently this model only simulates the latency introduced by software and does not consider realistic DMA (Direct Memory Access) or other hardware components which may introduce additional latencies. Their measurement setup was based on a Debian live Linux distribution and ixgbe NIC driver. It should be noted, that theses models are not part of the current ns-3 release at the state of this thesis and the instrumentation and development to achieve models which are applicable for several states of system utilization as presented in [113] is quite cumbersome.

In this work, a more simplified approach is used where the focus is on the estimation of a maximum number of hops possible for a single VoIP connection by including an appropriate processing delay in the network simulation. This processing delay is estimated for a single state of utilization from testbed measurements shown in Chapter 7.

Part I

System Design

Chapter 4

System Concept of the Disaster Recovery System

Contents

Section 2.1 showed an overview of research concepts and state-of-the-art Disaster Recovery Systems. The concept proposed in this work does not raise the claim to replace existing communication systems of governmental Emergency- and Rescue Services, but is meant to help volunteers and people in need. In this chapter, the system overview of the new DRS approach which is based on the mesh network amendment for WLAN devices IEEE 802.11s [15] is developed. The following sections are intended to clarify some of the basic questions which are posed on the DRS, basically the coverage of the system, how many nodes are needed to establish a sufficient network and what is the lifetime of that system. These questions are addressed independently of the used protocols by system level simulations. A performance analysis by network simulations of the proposed system and studying the influence of the used protocol stack is detailed in Part II. The lifetime of the system will be analyzed by Monte Carlo simulations.

Chapter 5 is dedicated to the development of an algorithm for the lifetime enhancement of the disaster recovery network as well as to the evaluation of the expected lifetime enhancement provided by the algorithm.

4.1 System Overview and Requirements

The DRS proposed in this work has the following system, hardware and service requirements:

1. System requirements:

   - Instantaneous availability of the system within the first hours after the occurrence of a disaster.

   - Formation of the system by standard user equipment.

   - Access to the system with standard user equipment.

   - Enhancement of the lifetime of the DRS while still providing sufficient coverage and network performance.

2. Hardware requirements:

   - WLAN equipment with IEEE 802.11s mesh network capability.

   - Low power consumption of WLAN devices.

3. Service requirements:

   - Communication of rescuers and victims by a VoIP connection.

   - Exchange of text or voice messages.

   - Internet connection, if MSTAs with network access and gateway capability are still available.

In Section 2.1.2 the main approaches for public disaster recovery systems have been discussed, covering different concepts to establish the network itself as well as possible applications providing the required services over the network. The development of a suitable application on top of the used lower-layer protocol stack of the DRS is out of the scope of this work. FireChat [32] and the smarter-project [14] are first existing examples of such applications, which are intended to run on networks that are mainly based on user equipment, e. g. smart phones. As has been detailed in Section 2.1.2 lifetime of such networks based on user equipment is considered one of the key open issues in the corresponding research work [14] and [32]. Including cars and home appliances with larger batteries could partially solve this issue and provide a better transport network for this kind of applications. Including these kind of devices, however is a difficult task, when no mesh network approach is used for the transport network. Both mentioned

App's thereby don't rely on the IEEE 802.11 mesh network amendment. FireChat is based on the IEEE 802.11 ad-hoc mode whereas Serval Mesh which is used in the smarter-project is based on the AP and Client mode. Both networking approaches need an additional forwarding protocol on higher layers which is not the case for the IEEE 802.11s mesh mode operating on layer 2 only. In terms of the interoperability of different types of user equipment (e. g. smart phones and laptops) and MSTAs in home appliances and cars the amendment for wireless mesh networks IEEE 802.11s seems to be most suitable. The system design for the DRS network therefore focusses on mesh networks only.

Even when restricting the system design to mesh networks only, several network topologies have to be considered. Figure 4.1 shows the topologies of the two fundamental scenarios that can occur. On one hand the DRS can be formed by user equipment only, i. e. by smart phones and laptops, on the other hand the network essentially is formed by MSTAs which are provided by remaining Wi-Fi devices in home appliances and cars in the disaster region. In both cases people in need can gain access to the network with their smart phones or laptops which are also part of the network. All stations, i. e. the user equipment as well as the additional stations, e. g. in home appliances or in cars, are assumed to be capable of the amendment for wireless mesh networks IEEE 802.11s. The on-demand mode of the Hybrid Wireless Mesh Protocol (HWMP) which is defined as the default path selection protocol in IEEE 802.11s as described in Section 2.2.3.2 thereby will be used. A further analysis of HWMP on the DRS will be shown by network simulations in Part II. The two Mesh Basic Service Sets (MBSSs) shown in Fig. 4.1 may also interconnect with each other through the DS as described in Section 2.2 which allows the communication of MSTAs of different MBSSs.

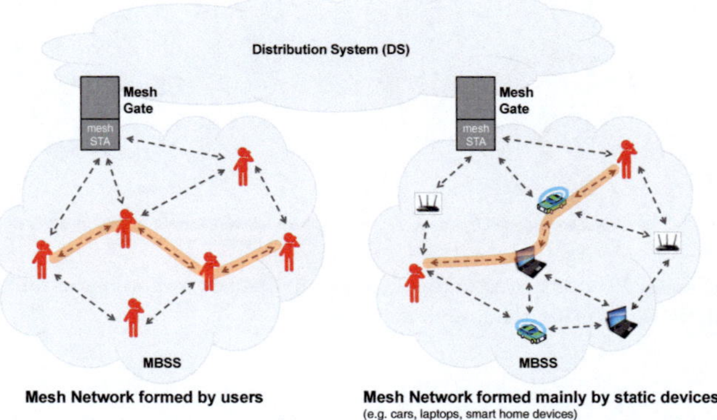

Figure 4.1: Disaster Recovery System topologies for the two fundamental user scenarios. The DRS can be either based on connected MSTAs in home appliances, cars and standard user equipment, e. g. smart phones and laptops, or only on user MSTAs.

The disaster recovery network is mainly static with low mobility because of its formation by the mainly static MSTAs still functional in the disaster area and because of the low mobility of the users in the disaster area. The static MSTAs are needed to form the network while guaranteeing a certain coverage for the users. If the number of static MSTAs is low the coverage has to be guaranteed mainly by the user equipment and thus is almost completely under the responsibility of the users. The drawback of a system formed only by the equipment of the users, however, is the lower system lifetime because of the lower battery capacity of the user equipment as well as a less predictable behavior of the network due to mobility of the users. Therefore this work will focus essentially on systems formed by static devices, which appears to be a very probable one, especially due to the strongly increasing presence of Wi-Fi devices in home appliances and even in cars.

If one of the connected stations still has connection with another communication network such as satellite or mobile communication networks (e. g. GSM, UMTS, LTE and WiMAX), a connection with other regions and internet access might be possible (see Fig. 4.2).

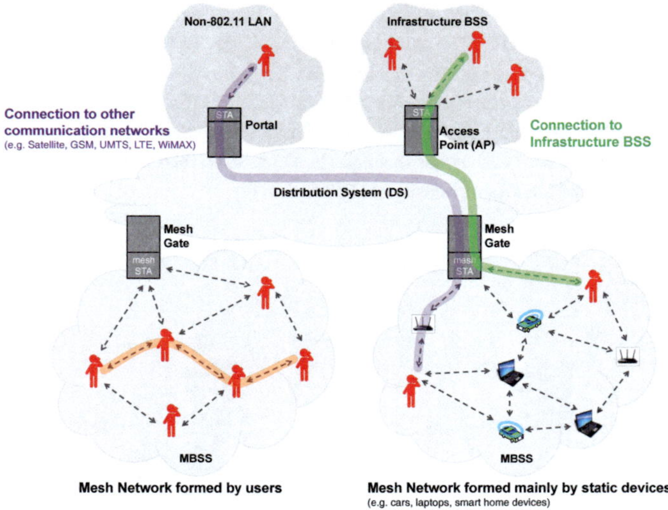

Figure 4.2: Disaster Recovery System topologies with different user scenarios for the connection to networks other than mesh networks.

A collocated device may exist which combines two different entities (MSTA and Portal) which allows the communication to other networks using the Portal. Additionally, STAs without mesh capabilities which are associated to an AP may communicate via a Mesh Gate. This might be possible by devices which combine two different entities (MSTA and Access Point). Figure 4.2

shows the topologies of different user scenarios with the connection to networks other than mesh networks. If all static devices combine the functionalities of a Mesh Gate and an Access Point in a single device allowing the communication of associated STAs, the mesh network could also be seen as a backbone communication network for the user STAs without mesh capabilities.

The simulations carried out in the following sections onwards are based on a single disaster recovery mesh network as shown by one of the MBSS in Fig. 4.1. However, the results for the coverage of the mesh network will also be valid for additional randomly placed STAs in the network (without mesh capabilities) which are associated to APs. The STAs are located inside the covered area, but do not enhance the overall covered area because these STAs are only connected to one of the APs and not to other STAs which could provide an enhancement of the coverage by relaying messages.

The most important hardware requirement to be mentioned at this stage is the support of the IEEE 802.11s mesh mode by the equipment forming the DRS, where it has to be stated that this capability nowadays is not commonly implemented in WLAN devices. However, as IEEE 802.11s provides only extensions to the MAC-layer, but not to the PHY this seems not to be an issue and might be handled by software updates. As an example the open-source GNU/Linux distribution for embedded devices OpenWrt [16] already contains an IEEE 802.11s implementation.

Also the availability of WLAN in home appliances and cars is moderate at the time of this work, but tends to be increasing in the coming years by the idea of internet of things and the connected car. The functionality of a DRS mode based on wireless mesh networks in cars thereby might be regulated by law similarly to the in-vehicle emergency call (e-Call) which has been recently adopted by the European Parliament and standardized by ETSI.

Of course the lifetime of the system is dependent on the lifetime of each MSTA. The development of low-power communication equipment for the consumer market is a general topic. A low-power small size IEEE 802.11a/g/n Wireless LAN module has been developed during this work in the FP7 research project CHAMELEON (Project reference: 286510) and is further described in Section 7.1.2. This module will also be used as a reference for the analysis of the fundamental multihop performance using the open-source OpenWrt IEEE 802.11s implementation.

The coverage of the network is dependent on the number of available MSTAs and also on the coverage of each individual MSTA. The number of MSTAs needs to be high enough to cover the disaster area. However, a higher number of available MSTAs may lead to additional interference which has a bad impact on the link quality. Measurements throughout an urban area in Germany reported in [17] revealed that 1971 WLAN Routers were available in an area of $467.500 \, \mathrm{m}^2$. This would lead to about 4200 routers/km^2 in an urban area not including additional user equipment in this area.

The system coverage in a disaster area which is based on the available number of MSTAs and their transmission range will be further analyzed in Section 4.2 by state-of-the-art analytical calculations and Monte Carlo simulations carried out in Matlab.

The previously described requirements have to be provided over a maximum period of time. It has been already mentioned that personal wireless devices such as smart phones or laptops have a low battery lifetime between zero and 24 hours. WLAN interfaces available in cars or in any other battery powered electrical devices may have a much longer lifetime of several days because of the powerful batteries in these devices. The lifetime of the system will be analyzed in Section 4.3 by analyzing two different scenarios.

To enhance the lifetime of the WMN DRS an algorithm will be presented in Chapter 5 which extends the lifetime by switching off unneeded nodes and reactivate them later when needed. Performance improvements provided by this algorithm regarding lifetime of the network will by analyzed by Monte Carlo simulations carried out in Matlab.

4.2 System Coverage and Scalability

In the following sections the system coverage achievable by the DRS as well as its scalability with the size of the disaster area will be analyzed by analytical and simulative approaches. For these system level simulations the term node is used as synonym for a Mesh Station (MSTA).

4.2.1 System Coverage

The coverage of the whole wireless mesh network is dependent on the coverage of each single MSTA. In the following sections a simplified physical layer model is used for the calculation of the coverage provided by each single MSTA, namely the unit disk connection model as introduced in Section 3.1.2.1. In this section we alternatively call this model also the "fixed range model". Within the framework of the unit disk model or fixed range model a connection between two MSTAs is possible only if

$$\|x_i - x_j\| = d_r \leq r_0 \tag{4-1}$$

for a fixed transmission range r_0 of nodes i and j. The transmission range r_0 thereby will be calculated as shown by the link budget calculation in Section 3.1.1.3 using the ITU Urban Micro cell (UMi) path loss model which, following the analysis in Section 3.1.1.1, seems to be most appropriate in an urban disaster area at the considered frequency of 5.5 GHz. The random variations of the channel due to multipath effects or shadowing thereby are taken into account by adding suitable margins given by the path loss model [54, Tab. A1-7]. This results in an estimation for the covered area with associated probability. Table 4.1 shows the transmit parameters used for the estimation of the transmission range with a carrier frequency of 5.5 GHz. For all simulations the most robust modulation BPSK with coding rate 1/2 will be used which results in the highest possible transmission range.

Table 4.1: Transmit parameters for the calculation of the transmission range r_0.

Carrier Frequency	5.5 GHz
Tx power in dBm	20
Tx Antenna gain in dBi	5
Rx Sensitivity in dBm	-92
Modulation	BPSK coding rate 1/2
Rx Antenna gain in dBi	5
Shadowing Margin in dB	-5.1
Multipath Fading in dB	-9.8
Transmission Range in m	**60**

According to Section 3.1.1.2 an allowed path loss of 107.1 dB for 90% coverage can be calculated with these parameters for the ITU UMi NLOS model. The resulting transmission range of $r_0 = 60\,\mathrm{m}$ will be used in the following simulations which indicates that if the distance d_r between two nodes is less or equal than r_0 ($d_r \leq r_0$) the nodes are connected and a transmission is always feasible (see Fig. 4.3). In the example of Fig. 4.3 nodes n_0 to n_6 set up a connected cluster within which a transmission is always feasible. Node n_7, however, is isolated because it is out of range of all other nodes and a transmission over n_7 would never be possible in a static network without mobility.

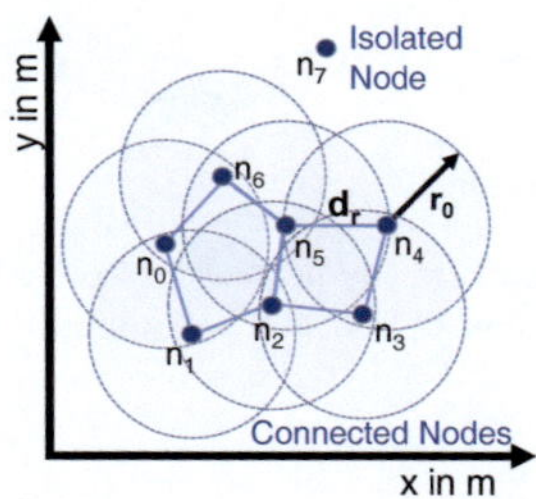

Figure 4.3: Example of the connectivity of nodes n_0 to n_7 using the unit disk connection model.

4.2.2 Analytical Estimation of the Node Density for the DRS

In Section 3.1.2.3 an analytical expression has been presented which allows the determination of the required transmission range r_0 that creates, for a given node density $\rho = N/A$ with N nodes and area A, an almost surely 1–connected network, i.e. a network without isolated nodes.

Likewise the required node density for a specified transmission range r_0 can be calculated. A connectivity of the network with $k = 1$ means that for each pair of nodes there exists at least one path connecting the two nodes of the pair. A redundancy in the network in this case is not given and if one of the nodes fails this may lead to a significant reduction of coverage. Therefore the resulting node density allows for the calculation of the minimum number of nodes N needed in a certain area A. As already derived in Section 3.1.2.3 Fig. 4.4 shows the probability $P(d_{\min} \geq 1)$ that no node is isolated for an area $A = 250\,\mathrm{m} \cdot 250\,\mathrm{m}$ as a function of the number of nodes N in the area A and the transmission range r_0 of each node according to Eq. (4-2) where $d_{\min}$ denotes the minimum node degree, i. e. the minimum number of connected neighbors that each node in the network at least has to have.

$$P(d_{\min} \geq 1) = (1 - e^{-\rho \pi r_0^2})^N \tag{4-2}$$

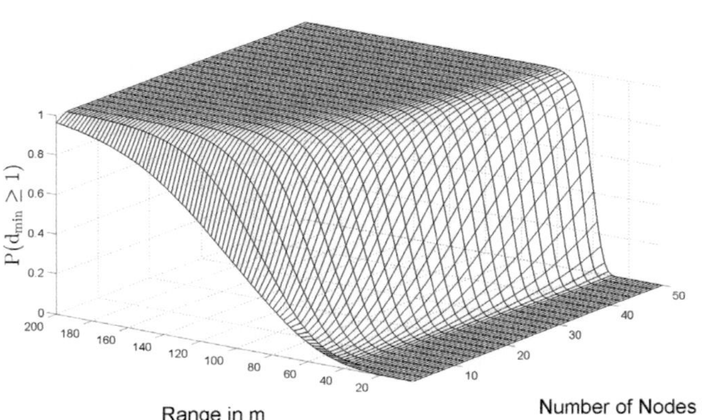

Figure 4.4: Probability that no node in the network is isolated as function of the transmission range r_0 and the number of nodes N according to Eq. (4-2), $A = 250\,\mathrm{m} \cdot 250\,\mathrm{m}$.

In the following we will denote with the term neighbor only those neighboring nodes to a given node, which are closer to the given node than the threshold distance r_0 and to which therefore a connection is possible.

The size of the area A has been chosen in order to reduce the computational overhead in the simulations of the following sections by a reduced number of nodes. Eq. (4-3) thereby shows the expected number of neighbors $E\{d\}$ of a node according to [64, Eq. 16] which equals the average node degree d_{mean} of the resulting graph.

$$E\{d\} = \rho \pi r_0^{\,2} \tag{4-3}$$

The scalability of this approach to larger areas will be further investigated. To this end, we next look on different sizes of the area A.

Table 4.2 shows the estimation for the required number of nodes according to Eq. (4-2) for different sizes of the areas A and different probabilities $P(d_{min} \geq 1)$ together with the expected number of neighbors for each node as calculated by Eq. (4-3). Let's repeat that all results throughout the Chapters 4 and 5 are obtained, using the fixed range model as introduced at the beginning of this paragraph. Thus the calculations in Section 4.2.2 onwards are based on the assumption of a fixed transmission radius within which all neighboring nodes are connected. It should be noted that there is still a certain probability depending on r_0 that a connection is not working which, however, is out of scope of this model with fixed radius. In Section 3.1.1.2 it was shown that a coverage of 90% can be reached for the chosen path loss model with a transmission range $r_0 = 60\,\text{m}$ of a single MSTA.

Table 4.2: Required number of nodes and expected number of neighbors for areas of different sizes A and probabilities $P(d_{min} \geq 1)$ in %.

Area $A = 250\,\text{m} \cdot 250\,\text{m}$		
$P(d_{min} \geq 1)$ in %	90	99
N (number of nodes) required	32	47
Node density ρ in nodes/km^2	512	752
r_0 in m	60	60
Expected number of neighbors E $\{d\}$	5.79	8.5
Area $A = 1000\,\text{m} \cdot 1000\,\text{m}$		
$P(d_{min} \geq 1)$ in %	90	99
N (number of nodes)	789	1019
Node density ρ in nodes/km^2	789	1019
r_0 in m	60	60
Expected number of neighbors E $\{d\}$	8.92	11.52

Table 4.2 shows that with a fixed transmission range r_0 of 60 m and 47 nodes almost all nodes are connected, i.e. $P(d_{min} \geq 1) = 99\%$, in the specified area $A = 250\,\text{m} \cdot 250\,\text{m}$ (see Fig. 4.4). Assuming a larger area of $A = 1000\,\text{m} \cdot 1000\,\text{m}$ about 1000 nodes are needed in order to reach this probability for an almost surely connected network. The scalability in terms of the node density ρ and the expected number of neighbors is also shown in Tab. 4.2 by comparing the results of the two different sizes of the area A. It can be observed, that both the node density ρ as well as the expected number of neighbors E $\{d\}$ increase with increasing area A if we stick to the same connectivity, i.e. $P(d_{min} \geq 1) = 90\%$. This is somehow counterintuitive.

We thus have to investigate this issue in more detail. Figure 4.5 shows how the number of nodes required for a given probability $P(d_{min} \geq 1)$ for a fully connected network scales for the analytical approach given by Eq. (4-2) by successively doubling the area A for $r_0 = 60\,\text{m}$ and initial value of the area $A = 250\,\text{m} \cdot 250\,\text{m}$ up to an area with surface $A_6 = 32A$. Looking at a probability of $P(d_{min} \geq 1) = 90\%$, $r_0 = 60\,\text{m}$ and area $A = 250\,\text{m} \cdot 250\,\text{m}$ a scaling factor for the number of nodes of about 2.18 can be determined which shows that the number of nodes is more than doubled when doubling the area A. This factor holds also for different sizes of the initial area A and transmission radii r_0 with minor differences (see Tab. 4.3). The scaling factors are determined by an exponential curve fitting of the scaling function assuming the form $f(x) = a \cdot e^{bx}$. The fitting is carried out for the calculated results of the probabilities $P(d_{\min} \geq 1) = 90\%$ of the areas $A_1 = A, A_2 = 2A, A_3 = 4A, A_4 = 8A, A_5 = A16A, A_6 = 32A$ for $x = 1, 2, 3, 4, 5, 6$. The scaling factor then equals to e^b.

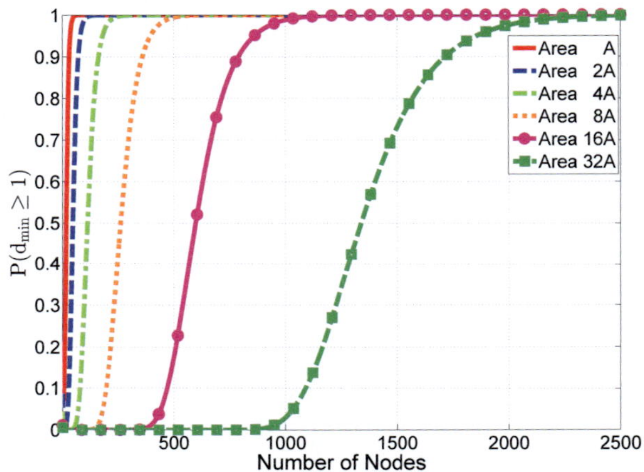

Figure 4.5: Probability that no node in the network is isolated as function of the number of nodes N with transmission range $r_0 = 60\,\text{m}$ for different sizes of the area A up to $32A$, starting at $A = 250\,\text{m} \cdot 250\,\text{m}$.

Table 4.3 shows also the scaling factors for an initial size of the area $A = 1000\,\text{m} \cdot 1000\,\text{m}$ up to an area of $32A$. The system doesn't scale exactly when increasing the area A to larger areas. The reason is that only the area A has been increased and not the transmission radius r_0. This, however, changes the graphical properties of the graph which leads to the result that more nodes are needed to achieve the same probability $P(d_{\min} \geq 1) = 90\%$ because of the lower proportion of r_0^2/A. Of course the scaling factors remain the same when doubling r_0 and quadrupling A according to Eq. (4-2) which means that the graphical properties remain the same. This is also shown by the first and last row of Tab. 4.3 where r_0 is quadrupled and A is the sixteenfold of the initial value with the resulting scaling factor of 2.153. If the transmission range r_0 remains unchanged while upscaling the area A more nodes are needed to achieve a similar probability.

From these results it is clear, that for increasing the size of the area A the node density required for a fully connected network will tend to infinity for continuously growing size of the area A. As already outlined in Section 3.1.2.3, this analytical result can be easily understood, however, is not suitable when looking on real networks where a small number of isolated nodes can

Table 4.3: Scaling factor for the required number of nodes to cover different areas using different transmission radii for $P(d_{\min} \geq 1) = 90\%$ considering for each row area sizes from A to $32A$ starting at the indicated initial value of A.

Area (initial value)	Transmission Range	Scaling Factor
$A = 250\,\mathrm{m} \cdot 250\,\mathrm{m}$	$r_0 = 30\,\mathrm{m}$	2.153
$A = 250\,\mathrm{m} \cdot 250\,\mathrm{m}$	$r_0 = 60\,\mathrm{m}$	2.183
$A = 250\,\mathrm{m} \cdot 250\,\mathrm{m}$	$r_0 = 120\,\mathrm{m}$	2.222
$A = 1000\,\mathrm{m} \cdot 1000\,\mathrm{m}$	$r_0 = 30\,\mathrm{m}$	2.115
$A = 1000\,\mathrm{m} \cdot 1000\,\mathrm{m}$	$r_0 = 60\,\mathrm{m}$	2.131
$A = 1000\,\mathrm{m} \cdot 1000\,\mathrm{m}$	$r_0 = 120\,\mathrm{m}$	2.153

always be tolerated. Let's first have a closer look on why the required node density for a full connectivity of the network is monotonically increasing with the size of the covered area A. To this end we consider the example of the area $A_5 = 1000\,\mathrm{m} \cdot 1000\,\mathrm{m}$ in Fig. 4.5 into which the initial area $A = 250\,\mathrm{m} \cdot 250\,\mathrm{m}$ fits 16 times.

The analytical result of Eq. (4-2) is equivalent to the assumption of independent probabilities for a connected network in each of the sixteen areas of size $A = 250\,\mathrm{m} \cdot 250\,\mathrm{m}$. The required number of nodes for achieving a probability of $P(d_{\min} \geq 1) = 90\%$ within the area $A_5 = 1000\,\mathrm{m} \cdot 1000\,\mathrm{m}$ thus corresponds to the result of $P(d_{\min} \geq 1) = 0.9^{(1/16)} \cdot 100 = 99.3\%$ for each of the sixteen areas A of size $A = 250\,\mathrm{m} \cdot 250\,\mathrm{m}$. The probability $P(d_{\min} \geq 1) = 99.3\%$ for each of the areas of size $A = 250\,\mathrm{m} \cdot 250\,\mathrm{m}$ is achieved by 49 nodes which leads to $784\,\mathrm{nodes/km^2}$. This is almost the required node density for an area of $A = 1000\,\mathrm{m} \cdot 1000\,\mathrm{m}$ according to Tab. 4.2. The slightly higher node density that can be calculated by Eq. (4-2) and which is presented in Tab. 4.2 is easily explained by the fact, that the above reasoning does not take into account the nodes required for the connectivity between the 16 individual sub-areas of size $A = 250\,\mathrm{m} \cdot 250\,\mathrm{m}$ which of course requires some additional nodes.

This shows that the stringent requirement for full connectivity of the network $P(d_{\min} \geq 1)$ is not a suitable criterion for the scalability of the node density with increasing number of nodes. It is clear, that the proposed DRS allows for a number of isolated nodes if most of the disaster area is covered. What thus has to be considered instead is the area that is actually covered by the DRS and within which a full connectivity is provided. The DRS thereby can be assumed to be sufficient if a sufficient percentage of the entire area is covered by the DRS whereas full connectivity within the entire area does not provide a suitable criterion. The following sections will thus focus on the calculation of the portion of the total area that the DRS is able to cover, providing full connectivity in this portion of the area.

4.2.3 Simulation of the Exact Covered Area

The analytical results presented in the previous section are based on the calculation of an upper bound (Eq. (4-2)) and because of the border effect the actually required node density for achieving a certain connectivity is always higher in a finite rectangular area. In this section therefore first the probability of achieving a connected set of all nodes is estimated by a Monte Carlo simulation in Matlab. Second, and as mentioned at the end of the previous section, this simulation is extended in order to calculate the covered area (in percent of the total area) i. e. the area that is covered by the transmission radii of the connected nodes. Again an area[1] of $A = 250\,\mathrm{m} \cdot 250\,\mathrm{m}$ is used for the simulation. Within this area a certain amount of MSTAs (nodes) are randomly placed for each simulation run.

For each run of the Monte Carlo simulation the nodes are randomly placed in the area by using a discrete uniform distribution for the calculation of the coordinates of each node. In order to achieve stable results the number of runs is set to $M = 10.000$ for each simulation with set of nodes N. It is assumed that each node has connection to each of its neighbors if the distance to the neighboring node is less or equal than r_0. All nodes that have connection by one or more hops over neighboring nodes will set up a cluster within which communication is possible and the probability of a connected cluster comprising all nodes, i. e. leaving no isolated nodes in the network which corresponds to $P(d_{\min} \geq 1)_{\mathrm{sim}}$ can be obtained from the simulation. The flow chart in Fig. 4.6a shows the executed algorithm used for the simulation. Table 4.4 shows the result of the simulation for $P(d_{\min} \geq 1)_{\mathrm{sim\_edm}}$ using the Euclidean distance metric (edm) for different amounts of nodes.

For 47 nodes only a probability of $P(d_{\min} \geq 1)_{\mathrm{sim\_edm}} = 68\%$ can be achieved compared to the 99% when using the analytical calculation shown in Tab. 4.2, whereas about 100 nodes are needed to achieve a probability of 99% in the simulation.

Table 4.4: Simulation results for $P(d_{\min} \geq 1)_{\mathrm{sim}}$ in % for area $A = 250\,\mathrm{m} \cdot 250\,\mathrm{m}$ and transmission range $r_0 = 60\,\mathrm{m}$.

N (Number of Nodes)	20	30	40	44	47	50	60	65	100
$P(d_{\min} \geq 1)_{\mathrm{sim\_edm}}$	2	17	47	60	68	73	87	91	99
$P(d_{\min} \geq 1)_{\mathrm{sim\_tdm}}$	33	85	97	98.8	99.3	99.7	99.96	99.96	100

The difference arises because the analytical results presented in the previous section are based on the calculation of an upper bound excluding any boundary effects (Eq. (4-2)). By using a toroidal distance metric (tdm, see Section 3.1.2.2 and Appendix C) which is used to eliminate the border effects the simulation results $P(d_{\min} \geq 1)_{\mathrm{sim\_tdm}}$ approach this upper bound. Figure 4.7 shows the comparison between the results of the analytical calculation according to Eq. (4-2) and the simulations with usual Euclidean distance metric for $P(d_{\min} \geq 1)_{\mathrm{sim\_edm}}$ and toroidal distance metric for $P(d_{\min} \geq 1)_{\mathrm{sim\_tdm}}$ for ascending number of nodes, Area $A = 250\,\mathrm{m} \cdot 250\,\mathrm{m}$, radius $r_0 = 60\,\mathrm{m}$.

---

[1]This area covers a small street block in an urban area.

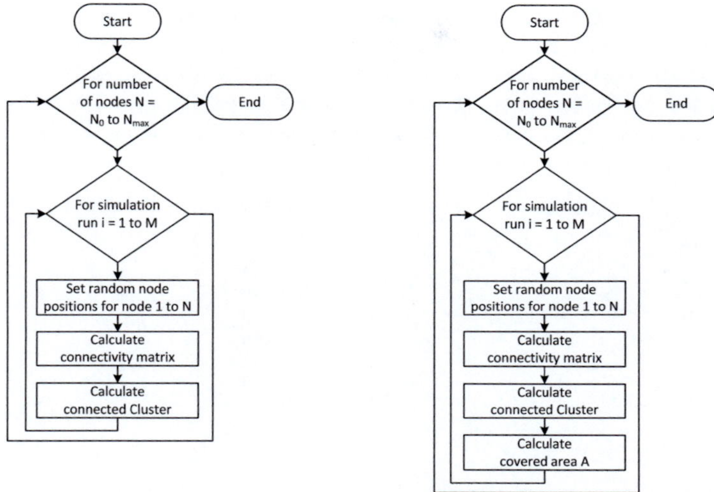

(a) Simulation of $P(d_{\min} \geq 1)_{\mathrm{sim}}$ for N_0 to $N_{\max}$ nodes for specified area A and transmission radius r_0.

(b) Extension of the algorithm shown in Fig. 4.6a to additionally calculate the covered area of the largest connected cluster with maximum contiguous area.

Figure 4.6: Flow chart of the Monte Carlo simulation to calculate the probability $P(d_{\min} \geq 1)_{\mathrm{sim}}$ for a connected network for N_0 to $N_{\max}$ nodes for specified area A and transmission radius r_0 and its extension allowing the calculation of the largest area covered by a connected cluster.

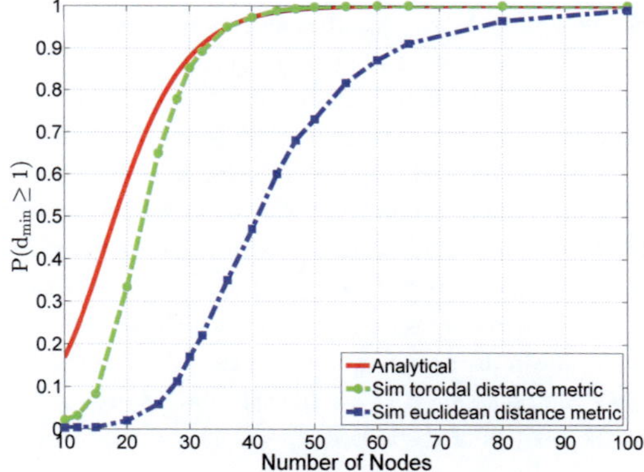

Figure 4.7: Simulation results for $P(d_{\min} \geq 1)$ for ascending number of nodes for analytical calculation and simulation with Euclidean and toroidal distance metric. Area $A = 250\,\mathrm{m} \cdot 250\,\mathrm{m}$, radius $r_0 = 60\,\mathrm{m}$.

In the use case of a DRS some isolated nodes are acceptable if most of the disaster area is still covered by the network. The Monte Carlo simulation is therefore extended to calculate the area covered by the connected cluster within the network that covers the largest contiguous area (see the flow chart in Fig. 4.6b). This area is calculated by evaluating the area of the cluster which covers the maximum contiguous area (isolated nodes and subclusters covering a smaller area are neglected), and is illustrated in Fig. 4.8.

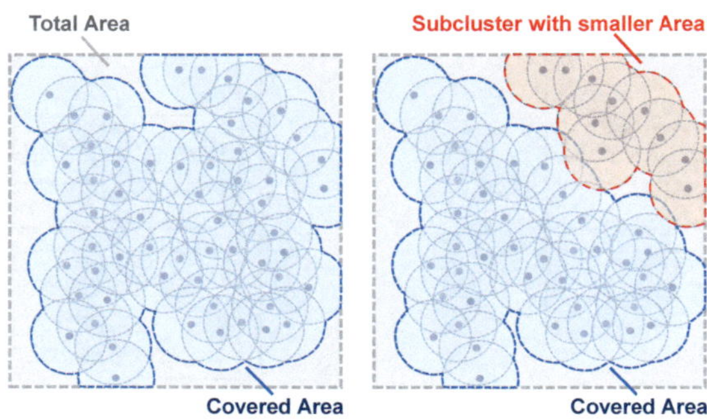

Figure 4.8: Examples illustrating the definition of the covered area within the total area A.

Thereby this maximum area is calculated by using a Monte Carlo integration with a number of 10.000 points uniformly distributed in the total area A. The covered area provided by the nodes is then estimated by the ratio of the number of points inside the contiguous area and the total number of distributed points.

Figure 4.9 shows the probability $P_{\text{covered area}}$ that an area of a certain size (given in percent of the total area) is covered as function of the size of the covered area for different amounts of nodes. The calculation of the covered area again is done by evaluating the covered area of all connected nodes of the largest clusters where the Euclidean distance metric is used.

Thereby $P_{\text{covered area}}$ denotes the probability for achieving a minimum covered area in percent of the total area with a given number of nodes N. Figure 4.9 thus shows that with a probability of $P_{\text{covered area}} = 90\%$ a covered area of at least 90% or higher of the $A = 250\,\text{m} \cdot 250\,\text{m}$ block can be achieved with 44 nodes at a transmission range of $r_0 = 60\,\text{m}$. With this number of nodes the probability of a fully connected network with no isolated nodes is only 60% when using the simulation with Euclidean distance metric (see Tab. 4.4). On the other hand, a covered area of 90% reasonably can be assumed to be sufficient for the proposed DRS. Isolated nodes will mainly occur at the borders of the area and thus most of the area is covered. A mobile user which is not connected to the network in most cases would be aware of finding himself at the edge of the covered area, e.g. by recognizing to be situated at the border of a formerly inhabited area, and thus most probably will try to walk into the covered area. This shows again that the probability of $P_{\text{covered area}}$ can be seen as a more appropriate metric than the

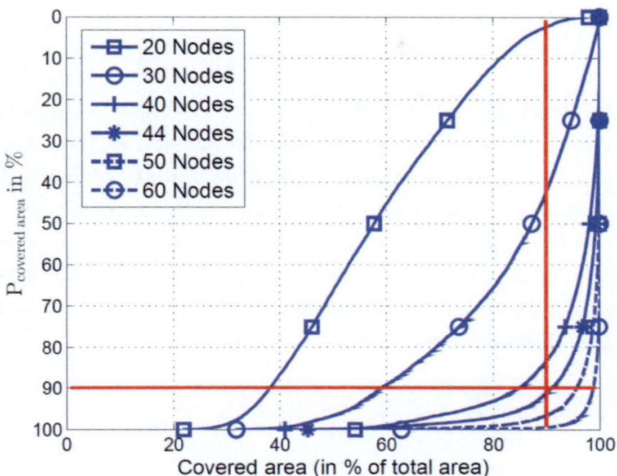

Figure 4.9: Probability of covered area in % of total area A, $A = 250\,\mathrm{m} \cdot 250\,\mathrm{m}$, radius $r_0 = 60\,\mathrm{m}$, simulation runs 10.000.

connectivity obtained by $P(d_{\min} \geq 1)$ and supports its use throughout the further analysis. Let us emphasize that for the size of the area of $A = 250\,\mathrm{m} \cdot 250\,\mathrm{m}$ the result for the required number of nodes for $P_{\text{covered area}} = 90\%$ is also rather close to the number of nodes for an almost surely connected network $P(d_{\min} \geq 1) > 90\%$ as given by the analytical results (see Tab. 4.2) where no border effects are present.

The result of 44 nodes required for $P_{\text{covered area}}$ (covered area $= 90\%$ of total area) $= 90\%$ finally leads to a required density of about 700 MSTAs/km^2 which is much less than the actual number of 4216 Routers/km^2 that has been revealed in the measurements in [17].

Table 4.5 shows the mean number of neighbor nodes NoN$_{\text{sim}}$ obtained from the simulations when using Euclidean and toroidal distance metric, area $A = 250\,\mathrm{m} \cdot 250\,\mathrm{m}$, transmission range $r_0 = 60\,\mathrm{m}$ and different number of nodes N. Again the result with 47 nodes and toroidal distance metric shows comparable results to the analytical results shown in Tab. 4.2.

Table 4.5: Simulation results for the mean number of neighbor nodes NoN$_{\text{sim}}$ when using Euclidean and toroidal distance metric, area $A = 250\,\mathrm{m} \cdot 250\,\mathrm{m}$ and different number of nodes N.

N (Number of Nodes)	20	30	40	44	47	50	60	65	100
NoN$_{\text{sim_edm}}$	3.26	4.52	5.83	6.37	6.79	7.21	8.62	9.34	14.40
NoN$_{\text{sim_tdm}}$	3.64	5.28	7.05	7.77	8.31	8.86	10.66	11.56	17.87

The expected mean covered area in percent of the total area $E_{\text{covered area}}$ can be calculated by averaging the ratio of the covered areas $A_{\text{covered area}}(i)$ to the total area $A_{\text{total area}}$ over all simulation runs M.

$$E_{\text{covered area}} = \frac{1}{M} \sum_{i=1}^{M} \frac{A_{\text{covered area}}(i)}{A_{\text{total area}}} \tag{4-4}$$

When using the Euclidean distance metric the mean covered area in percent of the total area for 44 nodes thereby is $E_{\text{covered area edm}} = 96.8\%$. Assuming a static network, the expected covered area $E_{\text{covered area edm}}$ also denotes the expected value of the probability for an additional randomly placed node (e. g. a user node) to be located inside this covered area $A_{\text{covered area}}$ and therefore also to be connected to the cluster. The probability for an additional user node to be covered by the area when using 44 static nodes thus would also be 96.8%. Table 4.6 summarizes the results for $E_{\text{covered area}}$ for different number of nodes N and when using the toroidal or Euclidean distance metric.

Table 4.6: Simulation results for $E_{\text{covered area}}$ in % with area $A = 250\,\text{m} \cdot 250\,\text{m}$ and transmission range $r_0 = 60\,\text{m}$.

N (Number of Nodes)	20	30	35	40	44	50	60
$E_{\text{covered area edm}}$	58.9	82.7	90	94.6	96.8	98.5	99.5
$E_{\text{covered area tdm}}$	83.0	97.2	98.6	99.2	99.5	99.7	99.88

4.2.4 Simulation of the Exact Covered Area in a Street Topology

In the previous section the simulations were carried out by simulating the MSTAs as a set of nodes randomly placed on a metric grid within the whole area A. In this section the nodes will be placed on a street topology reflecting the scenario where the DRS is established by cars equipped with MSTAs, which are parking along the streets. Street topologies, however, are different in rural and urban areas, vary from city to city and also from region to region in different continents. In Europe e. g., there is usually a clear division between the historical center and the more recent suburbs. In [115] a quantitative method to classify cities according to their street pattern is presented. In their work four large families of cities have been identified by different abundances of blocks of a certain area and shape based on the analysis of a set of 131 cities in the world. Most European and American cities in their sample, fall in their own sub-category, however, highlighting quantitatively the differences between the typical layouts of cities in both regions. It is also shown with the example of New York, that the fingerprint of a city can be seen as the sum of the different neighbourhoods inside this city as the topology can vary inside a city.

Group 1 (comprising Buenos Aires, Argentina only) essentially has blocks of medium size with shapes that are dominated by the square shape and regular rectangles. Small areas are almost exclusively squares. Group 2 comprises cities with a dominant fraction of small blocks with shapes broadly distributed. Athens, Greece is a representative example of group 2. Group 3 is similar to group 2 in terms of the diversity of shapes but is more balanced in terms of areas, with a slight predominance of medium size blocks. New Orleans, USA is stated as example of group 3. Group 4 contains for this dataset the interesting example of Mogadishu, Somalia which shows essentially small, square-shaped blocks, together with a small fraction of small rectangles. It is stated that, the characteristic of group 3 (various shapes with larger areas) dominates among cities in the world. All North American cities they have analyzed (except Vancouver, Canada) are part of group 3, as well as all European cities except Athens, which belongs to group 2 [115].

4.2.4.1 Manhattan grid

In the following section two simulations are carried out based on two different street topologies which rely on a simple Manhattan grid topology. Two street patterns may have the same topology, but blocks could be of very different size leading to different structures of the cities. The two topologies used here which represent two different topologies are Manhattan, New York USA and Mogadishu, Somalia. Figure 4.10 shows one of the topologies with 500 nodes marked as red circles. The topology represents a part in Manhatten with an area of $1000\,\text{m} \cdot 1000\,\text{m}$ from W15th Street to W27th Street and 8th Ave to 5th Ave. The topology is composed of medium size rectangles with dimensions of 75 m in vertical direction and 275 m in horizontal direction starting from the center of the streets. The streets which are represented by two delimiting lines have a width of 20 m for those in vertical direction and of 10 m for those in horizontal direction.

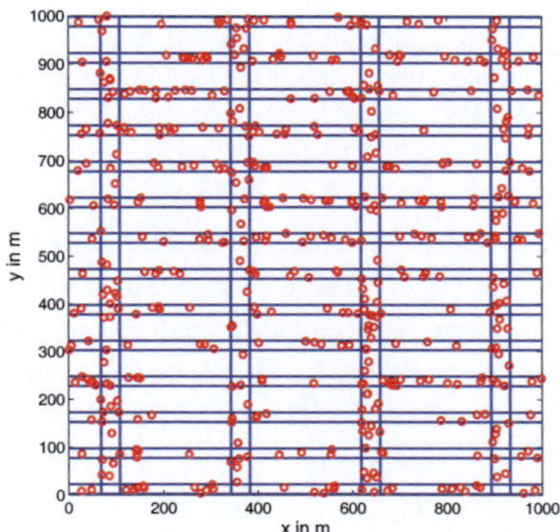

Figure 4.10: 500 nodes placed in the area $A = 1000\,\text{m} \cdot 1000\,\text{m}$ with Manhattan topology, reflecting Manhattan, New York.

The algorithm for randomly placing the nodes works as follows: First, two uniformly distributed pseudorandom integers in the interval [0,1000] are generated, representing the coordinates of a node in the topology. Then it is checked whether these coordinates are within the inner area delimited by one of the pairs of straight lines which are representing the streets. If this point is not within one of the areas identified as the streets a new set of coordinates is generated. This is repeated until the coordinates for the desired number of nodes for the topology are calculated.

Figure 4.11 shows the probability $P_{\text{covered area}}$ that an area of a certain size (given in percent of the total area) is covered as function of the size of the covered area for 100 up to 1000 distributed nodes in the Manhattan topology, reflecting Manhattan, New York. With a probability of $P_{\text{covered area}} = 90\%$ a covered area of approximately 90% of the $A = 1000\,\text{m} \cdot 1000\,\text{m}$ block can be achieved with about 750 nodes at a transmission range of $r_0 = 60\,\text{m}$.

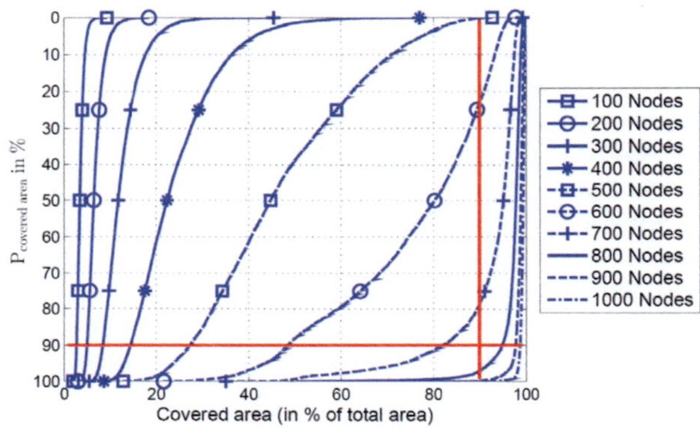

Figure 4.11: Probability of covered area in % of total area A, $A = 1000\,\text{m} \cdot 1000\,\text{m}$, radius $r_0 = 60\,\text{m}$, simulation runs 1.000. Between 100 and 1000 nodes, randomly placed in Manhattan topology, reflecting Manhattan, New York.

Figure 4.12 shows an example topology with 500 nodes realized following the topology representing the structure in Mogadischu. The topology represents a part near the Aden Adde International Airport. The topology is composed of small rectangles with size of 55 m in vertical direction and 110 m in horizontal direction starting from the center of the streets. The widths of the streets are 6 m in both directions.

Figure 4.13 shows the probability $P_{\text{covered area}}$ that an area of a certain size (given in percent of the total area) is covered as function of the size of the covered area for 100 up to 1000 distributed nodes in the Mogadischu topology. Compared to the Manhattan topology less than 600 nodes are needed to cover the same area of 90% with probability $P_{\text{covered area}} = 90\%$.

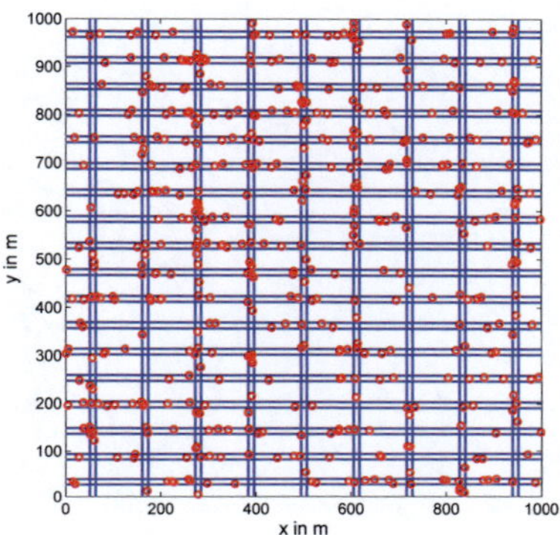

Figure 4.12: 500 nodes placed in the area $A = 1000\,\text{m} \cdot 1000\,\text{m}$ with Manhattan topology, reflecting Mogadischu.

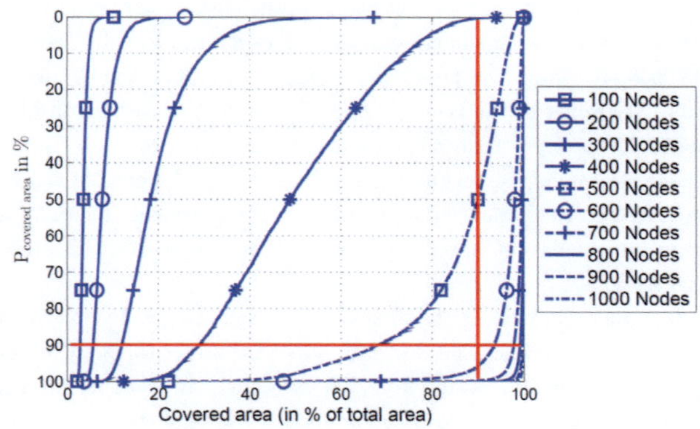

Figure 4.13: Probability of covered area in % of total area A, $A = 1000\,\text{m} \cdot 1000\,\text{m}$, radius $r_0 = 60\,\text{m}$, simulation runs 1.000. Between 100 and 1000 nodes, randomly placed in Manhattan topology, reflecting Mogadischu.

4.2.4.2 Rural street topology

The following simulations are based on the topology of the small village Illerrieden (nearby Ulm, Germany) with 2280 inhabitants, in the constituent state Baden-Wuerttemberg, which represents a typical rural street topology in Germany. This assumption contains a large part of subjectivity, however, the simulation and comparison of several rural areas in order to get a more scientific classification would be too complex at this stage. Figure 4.14 shows the street topology.

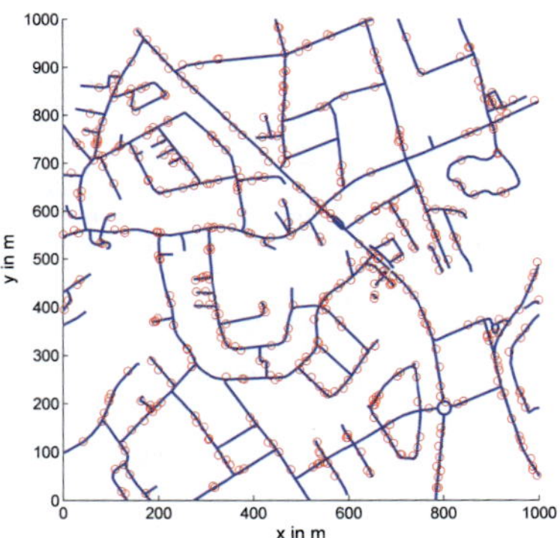

Figure 4.14: 500 nodes placed in the area $A = 1000\,\text{m} \cdot 1000\,\text{m}$ with rural street topology.

The topology of a $1000\,\text{m} \cdot 1000\,\text{m}$ area in the centre of the village has been extracted from Open Street Map [116] for the following simulations. The zero-point in Fig. 4.14 is at the geographic coordinates (48.2659,10.0466). The population density per square km in this region is about 1880 inhabitants. According to the Federal Office for Motor Vehicles (Kraftfahrt-Bundesamt) in Germany about 550 cars per 1000 inhabitants are usually expected in Germany. However, this number might be even higher in rural areas than in cities because of some lack in public transportation. Based on the number of inhabitants in the considered region of the village and using the average car population, about 1000 cars can be expected in this area and therefore up to 1000 mesh stations.

The algorithm for randomly placing the nodes has the following differences to the algorithm explained in the previous section: The raw data set of the street topology is based on several small straight line segments as a result from the extracted raw data from Open Street Map [116]. Again, two uniformly distributed pseudorandom integers in the interval [0,1000] are generated, representing the coordinates of a node in the topology. Then it is checked whether these coordinates are at a maximum distance of $5\,\text{m}$ from one of the straight lines which represent street segments[2]. If this point is not nearby, i.e. at a maximum distance of $5\,\text{m}$ of one of the straight line segments, a new set of coordinates is generated. This is repeated until the coordinates for the desired number of nodes for the topology are calculated. In Fig. 4.14 500 nodes are shown in the topology marked as red circles. Once all coordinates are calculated, the simulation of the previous section is carried out in order to calculate the probability $P_{\text{covered area}}$ that an area of a certain size (in percent of the total area) is covered as function of the size of the covered area for 100 up to 1000 nodes which are distributed on the street topology.

Figure 4.15 shows the results of the simulation with the rural street topology. A covered area larger than 90% of the $A = 1000\,\text{m} \cdot 1000\,\text{m}$ block can never be achieved with a probability of $P_{\text{covered area}} = 90\%$. The reason is the larger blocksize delimited by the street segments.

---

[2]Thus the streets have a width of $10\,\text{m}$ including parking areas and driveways.

Depending on the transmission range which was set to $r_0 = 60\,\text{m}$ in this simulation not all parts in the grid can be covered which might be typical for this kind of topologies. Also some isolated nodes may exist at the border of the grid.

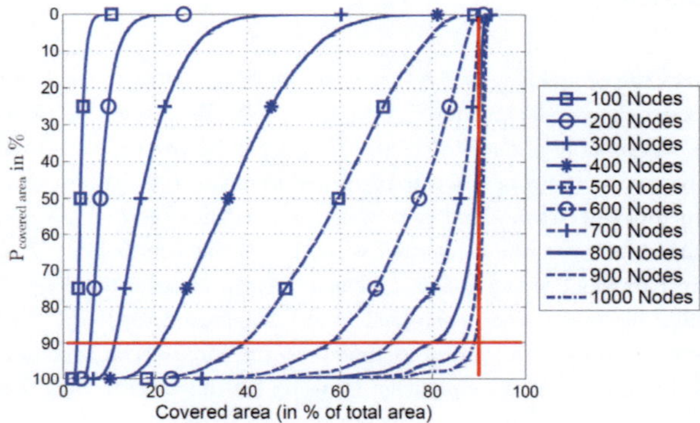

Figure 4.15: Probability of covered area in % of total area A, $A = 1000\,\text{m} \cdot 1000\,\text{m}$, radius $r_0 = 60\,\text{m}$, simulation runs 1000. Between 100 and 1000 nodes, randomly placed in rural street topology.

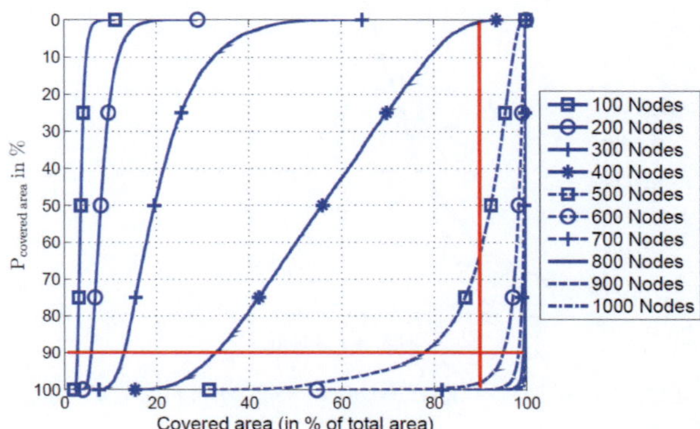

Figure 4.16: Probability of covered area in % of total area A, $A = 1000\,\text{m} \cdot 1000\,\text{m}$, radius $r_0 = 60\,\text{m}$, simulation runs 1000. Between 100 and 1000 nodes, randomly placed in the total area, i.e. without any restriction due to a street topology.

Finally, Fig. 4.16 shows the probability $P_{\text{covered area}}$ that an area of a certain size (in percent of the total area) is covered as function of the size of the covered area for 100 up to 1000 uniformly distributed nodes without street topology. With a probability of $P_{\text{covered area}} = 90\%$ a covered area of approximately 95% of the $A = 1000\,\text{m} \cdot 1000\,\text{m}$ block can be achieved with 600 nodes at a transmission range of $r_0 = 60\,\text{m}$.

4.2.4.3 Conclusion

Figure 4.17 shows the comparison of the three street topologies and of the topology with randomly placed nodes all over the area for $P_{\text{covered area}} = 90\%$. With the rural topology simulated in the previous section only about 90% of the area A can be covered. However, it should be noted that people may mainly stay close to roads and therefore 100% coverage is not necessary. The Manhattan topology reaches 100%, however, more nodes than in the random topology are needed in order to cover the same area with the specified probability $P_{\text{covered area}}$. The Mogadischu street topology shows a similar behavior as the random topology. The reason for this is the smaller block size when compared to the Manhattan topology. Assuming mainly disaster recovery scenarios with topologies similar to the Mogadischu topology the following simulations will be carried out, for simplicity, by using the random topology only.

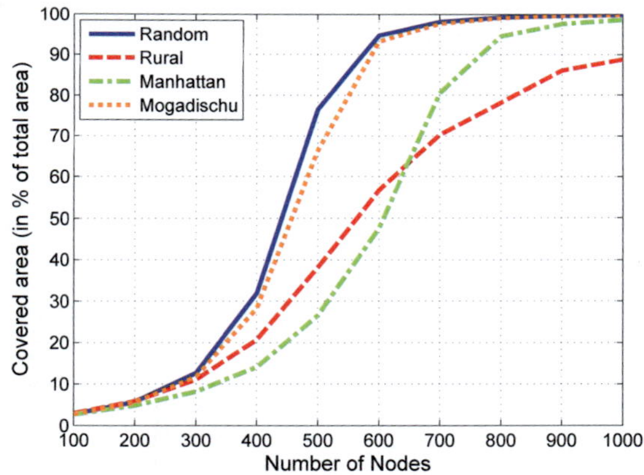

Figure 4.17: Comparison of the three street topologies for $P_{\text{covered area}} = 90\%$.

Tab. 4.7 shows the probability of a fully connected network $P(d_{\text{min}} \geq 1)_{\text{sim}}$ for different amounts of nodes with Euclidean and toroidal distance metric. For 1000 nodes only a probability of $P(d_{\text{min}} \geq 1)_{\text{sim\_edm}} = 75\%$ can be achieved with the random topology, however, $P(d_{\text{min}} \geq 1)_{\text{sim\_tdm}} = 98.6\%$ with toroidal distance metric. The Mogadischu street topology shows similar results with an even higher probability $P(d_{\text{min}} \geq 1)_{\text{sim\_edm}} = 85.6\%$ when applying the Euclidean distance metric. Due to the structured arrangement of the nodes the probability of an isolated node is less than in the simulation with random placement of the

Table 4.7: Simulation results of $P(d_{\min} \geq 1)_{\text{sim}}$ in % for different street topologies and with distance metrics.

Area $A = 1000\,\text{m} \cdot 1000\,\text{m}$							
N (Number of Nodes)	400	500	600	700	800	900	1000
$P(d_{\min} \geq 1)_{\text{sim\_edm}}$ Random	0	0.04	2.4	16.1	39.4	59.3	75.2
$P(d_{\min} \geq 1)_{\text{sim\_tdm}}$ Random	0	3.5	31.9	68.5	87.4	95.4	98.6
$P(d_{\min} \geq 1)_{\text{sim\_edm}}$ Rural	0	0	0	0.03	0.21	0.47	0.99
$P(d_{\min} \geq 1)_{\text{sim\_tdm}}$ Rural	0	0.02	2.02	13.35	33.3	53.1	68.8
$P(d_{\min} \geq 1)_{\text{sim\_edm}}$ Manhattan	0	0	0	0	0.02	0.2	1.6
$P(d_{\min} \geq 1)_{\text{sim\_tdm}}$ Manhattan	0	0	0	0	0.25	2.3	9.4
$P(d_{\min} \geq 1)_{\text{sim\_edm}}$ Mogadischu	0	0.05	4.1	24.9	52.4	73.7	85.6
$P(d_{\min} \geq 1)_{\text{sim\_tdm}}$ Mogadischu	0	1.5	24.3	60.8	83.4	94.2	97.7

nodes. The results of the simulations with the Manhattan and rural topologies show a very low probability of $P(d_{\min} \geq 1)_{\text{sim\_edm}} \approx 1\%$. The reason is that in the street topology almost always isolated nodes exist and the network can never be fully connected. Again, the probability of a fully connected network $P(d_{\min} \geq 1)_{\text{sim}}$ does not help much neither in the classification of the street topologies nor for answering the question how many nodes are actually needed to achieve a sufficient coverage.

Instead and as stated earlier a different evaluation methodology has to be envisaged: Tab. 4.8 shows the number of nodes in percent of the total number of nodes which are part of the biggest cluster for different street topologies denoted by $P_{\text{biggest cluster}}$.

Table 4.8: Simulation results for $P_{\text{biggest cluster}}$ in percent, i.e. the number of nodes in % of the total number of nodes which are part of the biggest cluster, for different street topologies and distance metrics.

Area $A = 1000\,\text{m} \cdot 1000\,\text{m}$							
N (Number of Nodes)	400	500	600	700	800	900	1000
$P_{\text{biggest cluster edm}}$ Random	55.88	89.1	97.56	99.22	99.71	99.87	99.94
$P_{\text{biggest cluster tdm}}$ Random	83.46	97.99	99.60	99.90	99.973	99.992	99.998
$P_{\text{biggest cluster edm}}$ Rural	42.89	66.30	82.72	91.18	95.28	97.16	98.02
$P_{\text{biggest cluster tdm}}$ Rural	49.65	78.02	92.46	97.43	98.99	99.54	99.76
$P_{\text{biggest cluster edm}}$ Manhattan	24.30	45.75	73.5	90.42	96.09	97.99	98.85
$P_{\text{biggest cluster tdm}}$ Manhattan	46.8	81.0	93.27	96.47	97.927	98.758	99.26
$P_{\text{biggest cluster edm}}$ Mogadischu	50.29	85.69	96.92	99.13	99.71	99.89	99.95
$P_{\text{biggest cluster tdm}}$ Mogadischu	74.05	96.57	99.3	99.83	99.95	99.986	99.995

Even though the probability of a fully connected network $P(d_{\min} \geq 1)_{\text{sim\_edm}} \approx 1\%$ being rather low for the Manhattan and the rural topology Tab. 4.8 shows, that still almost all nodes belong to the biggest cluster.

Finally, Tab. 4.9 shows the mean covered area and Tab. 4.10 the mean number of neighbor nodes for the topology with randomly placed nodes all over the area A.

Table 4.9: Simulation results for $E_{\text{covered area}}$ (in % of the total area) with area $A = 1000\,\text{m} \cdot 1000\,\text{m}$, transmission range r_0 and randomly placed nodes inside the area A for different number of Nodes N.

N (Number of Nodes)	400	500	600	700	800	900	1000
$E_{\text{covered area edm}}$	55.9	89.3	97.6	99.2	99.7	99.87	99.94
$E_{\text{covered area tdm}}$	83.0	97.6	99.4	99.8	99.92	99.96	99.98

Table 4.10: Simulation results for the mean number of neighbor nodes NoN_{sim} with Euclidean and toroidal distance metric, with area $A = 1000\,\text{m} \cdot 1000\,\text{m}$, transmission range r_0 and randomly placed nodes inside the area A for different number of nodes N.

N (Number of Nodes)	400	500	600	700	800	900	1000
Number of neighbors $\text{NoN}_{\text{sim\_edm}}$	4.63	5.51	6.50	7.53	8.58	9.65	10.71
Number of neighbors $\text{NoN}_{\text{sim\_tdm}}$	4.70	5.69	6.78	7.89	9.02	10.15	11.28

Note that $E_{\text{covered area}}$ as defined in Eq. (4-4) equals the probability for an arbitrary node, taken out of all nodes, to belong to the biggest cluster. The latter evaluation method, considering the biggest cluster, i.e. the cluster covering the largest contiguous area, will be used for the evaluation of the lifetime of the system in the following sections. Considering the result with randomly placed nodes in a rectangular area of $A = 1000\,\text{m} \cdot 1000\,\text{m}$, less than 600 nodes are needed to achieve a covered area of 90% of the total area with probability $P_{\text{covered area}} = 90\%$. This is half the number of nodes needed for an almost surely connected network as shown by the analytical calculations in Section 4.2.2, Tab. 4.2.

The mean number of neighbor nodes provides a good metric for the scalability of the network to larger areas. If we stick to 800 nodes the mean number of neighbor nodes with toroidal distance metric $\text{NoN}_{\text{sim\_tdm}}$ is about 9 nodes. The simulations with an area of $A = 250\,\text{m} \cdot 250\,\text{m}$ shown in Tab. 4.5, Section 4.2.3 showed 8.9 neighboring nodes with 50 nodes. As expected the number of neighboring nodes for a given coverage is independent of the size of the area when neglecting boundary effects and the number of nodes required for a given coverage scales with the size of the area, i.e. a factor of 16 for the example.

The results for $E_{\text{covered area}}$ (in percent of the total area) show also comparable results when using the Euclidean distance metric. However, the results obtained in Section 4.2.3 with an area of $A = 250\,\text{m} \cdot 250\,\text{m}$ show a lower mean covered area as well as a lower number of neighboring nodes, because of the boundary effects which are much more present at smaller areas.

The results shown so far for the number of nodes needed to cover a certain area when compared to the number of available nodes, e.g. the average car population or the number of available routers/km^2 as shown in [17], indicate that more nodes are expected to be available than needed to cover a certain area. Key for the DRS, however, is the lifetime of the system which will be further analyzed in the next section.

4.3 System Lifetime

4.3.1 Methodology

The lifetime of the DRS can be analyzed by extending the algorithm described in Section 4.2.3 by adding a certain lifetime to each node. When a node runs out of battery the node is shut down and the remaining covered area is calculated. Again the Euclidean distance metric is used for the calculation. Figure 4.18 gives an example: node n_4 runs out of battery resulting in two subclusters of nodes n_0 to n_3 and n_5 to n_7. Again the maximum contiguous area is calculated, in the example of Fig. 4.18 covered by nodes n_0 to n_3 only and resulting in a much smaller covered area for the remaining network lifetime.

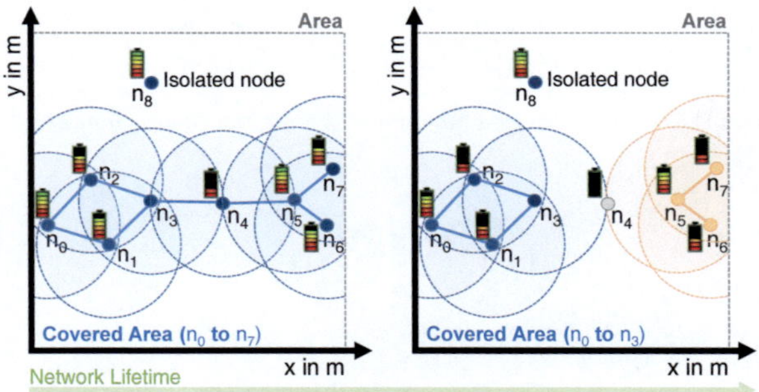

Figure 4.18: Illustration of the changes in the number of connected nodes during evolution of the network over time.

The lifetime of a DRS will be analyzed by evaluating two different scenarios. The first scenario will be based on battery powered "MESH Routers". The second scenario is based on the assumption that vehicles can be equipped with WLAN devices with mesh functionality according to IEEE 802.11s. In both scenarios the system can be seen as an individual backbone network for additional user nodes using those devices for the relaying of messages. Of course, the system as a whole forms a mesh network and all stations fall into the category MSTA. The distinction between "backbone" and "user" devices therefore does not reflect any differences in the type of network nodes, but merely reflects the fact, that the "left over devices" (mesh routers, cars) will not generate user traffic, but only traffic due to network management. Sources and sinks of traffic are only the user devices.

4.3.2 Scenario with Battery Powered Mesh Routers

Assuming the scenario with battery powered "MESH Routers", every MSTA is assumed to have a battery power lifetime in between four to six hours in the simulation. The battery lifetime of the individual MSTAs is modeled by using a Gaussian distribution with mean value of 5 hours and a standard deviation of 0.5 hours. The assumption thereby is that power consumption of the nodes is dominated by their standby state and independent of the traffic load which is assumed to be low. The simulation has been carried out with 102 MSTAs which leads according to the previous sections (Fig. 4.7) to an almost surely connected network at the initial state, before batteries start to discharge. Table 4.11 shows the parameters used for the simulation in Matlab.

Table 4.11: Simulation parameters of the scenario with battery powered mesh routers.

Area A	$250\,\text{m} \cdot 250\,\text{m}$
Transmission Range r_0	60 m
Number of MSTAs	102
Number of Monte Carlo runs	1000
Battery Lifetime	4 - 6 hours
Distance metric	Euclidean

Figure 4.19 shows the covered area as a function of the elapsed time in hours and the probability $P_{\text{covered area}}$ in %.

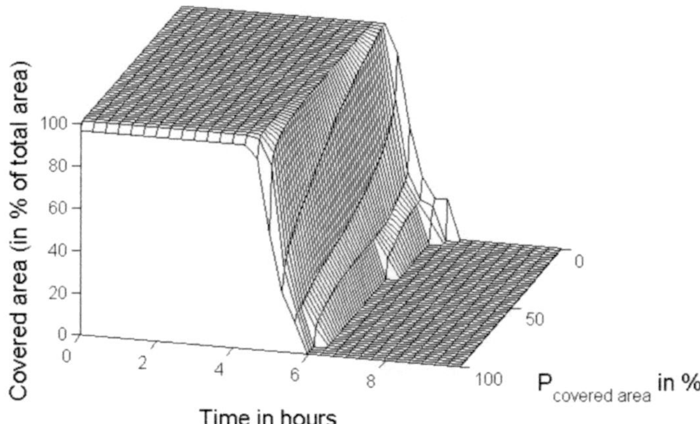

Figure 4.19: Covered area for $A = 250\,\text{m} \cdot 250\,\text{m}$, radius $r_0 = 60$ m, 102 nodes, 1000 Monte Carlo simulation runs and Euclidean distance metric of the scenario with battery powered mesh routers.

After approximately 5h the probability $P_{\text{covered area}}$ falls down to 90% if we stick to a required coverage of 90% of the total area for the DRS. After 6h most of the nodes are out of battery and the covered area is approaching zero. After more than six hours there would be almost no communication possible resulting in complete network outage of the DRS.

4.3.3 Scenario with Passenger Cars

The second scenario is based on WLAN devices in passenger cars which could provide a backbone network for people in need who are using their personal devices connected as MSTAs to the mesh network formed by the MSTAs in the cars. In this scenario the lifetime will be much longer compared to the previous section because of the powerful batteries in the cars and will be based on a simple calculation under the following assumptions. Modeling of battery lifetime of the cars will be based on the distribution of passenger cars according to segments, model series as well as typical sizes of the batteries in these cars. [117] reports the stock of passenger cars in Germany on January 1, 2016 by segment as well as by model series. Typical passenger car batteries have a capacity from 41 Ah up to 110 Ah. Table 4.12 shows the stock of passenger cars in Germany obtained from [117] for different segments and typical battery capacities for those segments which are defined by an interval of a uniform distribution. Figure 4.20 shows the histogram of the battery values shown in Table 4.12 which can be approximated by a Gaussian distribution with mean value of 70 Ah and standard deviation 15 Ah. Assuming a power consumption of 5 W this results in a Gaussian distribution for the lifetime of the MSTAs with mean value[3] $7 \cdot 24$ h and standard deviation $1, 5 \cdot 24$ h.

Table 4.12: Stock of passenger cars in Germany [117] and approximated battery capacities.

Car segments	Number of cars	Battery capacity distribution
Minis	3022272	uniform[40,60] Ah
Sub-Compact Class	8884452	uniform[52,70] Ah
Compact Class	11880160	uniform[60,80] Ah
Middle Class	7076772	uniform[70,95] Ah
Top Middle Class	2071543	uniform[80,105] Ah
Top Class	270178	uniform[85,105] Ah
SUV's	1813083	uniform[70,85] Ah
All-terrain	1864889	uniform[95,105] Ah
Sport Cars	839670	uniform[95,100] Ah
Minivans	2008076	uniform[60,80] Ah
Large capacity Vans	2076840	uniform[70,85] Ah
Utilities	1631625	uniform[60,80] Ah
Caravans	417297	uniform[80,110] Ah
Others	1214352	uniform[40,110] Ah

---

[3]Assuming 12 V car batteries the mean lifetime in days is calculated by 70 Ah/(5 W/12 V) / 24 h.

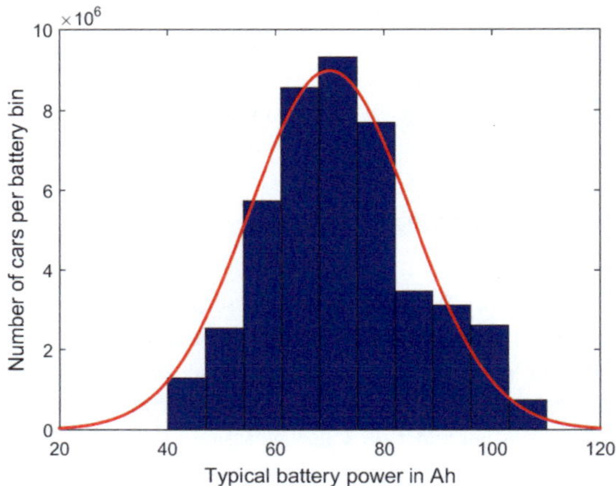

Figure 4.20: Statistic of battery power used for simulations.

Figure 4.21 shows the covered area as a function of the elapsed time in hours and the probability $P_{\text{covered area}}$ in %. After approximately 7.2 days the probability $P_{\text{covered area}}$ falls down to 90% if we stick to a required coverage of 90% of the total area for the DRS. After 9-10 days most of the nodes are out of battery and the covered area is approaching zero.

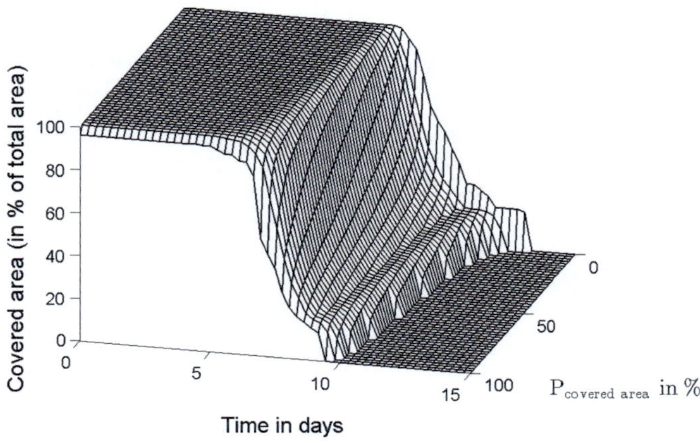

Figure 4.21: Covered area for $A = 250\,\text{m} \cdot 250\,\text{m}$, radius $r_0 = 60\,\text{m}$, 102 nodes, 1000 Monte Carlo simulation runs and Euclidean distance metric of the scenario with passenger cars.

4.3.4 Conclusion

Both examples show the dependence of the overall system lifetime on the battery lifetime at each individual Mesh Station. The second scenario based on car equipped MSTAs shows a much longer lifetime of several days because of the powerful batteries. However, nowadays in most cases cars are not yet equipped with WLAN devices. Also battery powered Mesh Routers are not very common. Therefore the lifetime will be mainly based on standard user equipment which has a similar lifetime of up to 24 hours. Therefore energy conservation is one of the essential requirements for a DRS. This requirement has to be reflected by the used protocol stack. A possible solution in order to enhance the lifetime of the system will be presented in the next chapter.

Chapter 5

Algorithm for Lifetime Enhancement of the DRS

Contents

Among the key considerations concerning the performance of a DRS are the number of nodes needed to set up a network in the specified area, i. e. the required node density, the coverage that can be achieved by a given node density as well as the lifetime of the system. For all those points, in addition, scalability to larger and smaller areas where the DRS should be established has to be considered. These constraints have been addressed in Chapter 4 showing that lifetime is a key issue. In this chapter an algorithm for the lifetime enhancement will be developed. The Lifetime Enhancement Algorithm (LEA) for Disaster Recovery System (LEA-DRS) proposed in this work requires no central control unit and allows to switch off devices that are not needed for realizing the required coverage in a first time period, especially in order to save the remaining energy in the batteries for a later usage when the initially switched on devices run out of battery. It will be shown that with the proposed algorithm for thinning out the initially available network the lifetime of a DRS can be considerably extended depending on the node density while keeping the required coverage of the disaster area.

An overview of different energy conservation mechanisms for WMNs based on results available in literature, the majority of which stem from considerations of Wireless Sensor Networks

(WSN), has been shown in Section 3.1.3. The type of DRS investigated in this work shows a number of key differences to Wireless Sensor Networks (WSN), which make the solutions proposed for WSN not applicable to the DRS. First, the disaster area is assumed to have a high node density, which is sufficient to set up the network. Second, the network has the ability to enable point-to-point communications between the individual nodes, which may forward the communication over several hops. Third, the throughput of the system is assumed to be rather low and thus the power consumption of the nodes occurs mainly during stand-by and not due to transmit or receive. One should bear in mind that those devices providing for connectivity of the mesh network on which the DRS is based can never be switched off, as there exists no central control unit that may command devices to switch back on when a communication should take place and communications can occur at any instance in time. Thus the lifetime of the DRS is essentially dependent on the stand-by lifetime of each individual node and on the remaining connectivity when nodes are going to be switched off due to lack of power. Finally, the nodes have to act individually in a distributed system as there is no central control unit available.

In order to achieve – respecting low energy consumption – a sufficient connectivity of the network one solution is the adaptive adjustment of transmission ranges (power adjustment). Alternatively a sufficient network planing in the pre-deployment phase may be used. These solutions may also correspond to a sufficient number of neighboring nodes which are achieved by the method. Both methods are not applicable for the DRS studied here, but it will be shown, that coverage can be guaranteed to a sufficient level by just respecting a sufficient number of neighbors for each station. This can be achieved by an approach which allows to shut down non-necessary nodes and to keep them for a later usage while still keeping network connectivity.

The principle presented in Section 4.3.1 for the calculation of the network lifetime will be turned into an algorithm that allows to shut down redundant nodes into a sleep mode and reactivate them later when needed. This allows for a considerable enhancement of the lifetime of the DRS for a given probability $P_{\text{covered area}}$. The algorithm is based on the following assumptions:

- The network is dense and more nodes than needed to cover about 90% of the area are present.

- Some of the nodes can be shut down into a power saving mode (PSM) while the connectivity and covered area of the remaining active nodes remain almost unchanged.

- These previously shut down nodes can be reactivated when needed to enhance the lifetime of the whole network while keeping its connectivity and covered area.

In a first approach the decision if a node is needed to guarantee connectivity will be based on the Euclidean distance d_r of the considered node to its neighboring nodes only and the shut down is decided by the nodes themselves. The assumption thereby is that if two nodes are close to each other, i.e. $d_r \leq d_{Th}$, where d_{Th} corresponds to a suitably chosen threshold distance, one of the nodes can be shut down without significant reduction in coverage. It should be noted that there is no guarantee that no isolated nodes can occur following this procedure. It is further assumed that the distance d_r can be estimated by the Received Signal Strength Indicator (RSSI) available as a PHY parameter and accessible to the processing units of the individual MSTAs.

In Section 5.1 first this method will be evaluated in a proof of concept by randomly shutting down nodes according to the threshold condition. Even though the algorithm is running deterministically, the Monte Carlo simulation reflects absence of a control instance.

Second, in Section 5.2 this approach will be extended to a distributed sleep-wake algorithm which is based on alternating sleep and awake states for each individual node. Finally, a variant of the algorithm, following an alternative approach, will be developed and compared to the previous one. This alternative approach consists in considering the number of neighboring nodes at each MSTA, instead of the RSSI as the criterion for the decision at each individual MSTA whether the station can be switched off. Thereby the number of neighboring stations is a parameter that is easily accessible by each MSTA through the peer link list according to IEEE 802.11s. Additionally, this parameter is considerably less ambiguous than the RSSI, because the number of peer links stems from execution of the IEEE 802.11s protocol only and has no vendor specific implementation as it is the case for the RSSI.

The results presented throughout Sections 5.1 and 5.2 have been published by the author of this thesis in [19] and [20].

5.1 Proof of Concept

5.1.1 Methodology

The first variant of the algorithm works as follows: if for a node n_1 there exists a neighboring node n_2 which is closer to n_1 than a given threshold distance d_{Th} (see Fig. 5.1a), n_1 decides to shut itself down into a power saving mode (Fig. 5.1b). Packets generally still can be routed over node n_2. The decision is made on basis of the threshold d_{Th} where out of each pair of nodes (n_i, n_j) at a distance $d_r \leq d_{Th}$ one node is shut down. With the coordinates (x_{i1}, x_{i2}) and (x_{j1}, x_{j2}) of the two nodes n_i and n_j the distance between these two nodes is calculated as $d_r(n_i \rightarrow n_j) = \sqrt{(x_{j1} - x_{i1})^2 + (x_{j2} - x_{i2})^2}$. The threshold distance d_{Th} is indicated by the inner circle inside the transmission range r_0 of the nodes in Fig. 5.1. This is repeated for every node in a sequential order. The algorithm relies on the assumption, that if two or more nodes are close to each other the additional nodes do not significantly enhance the overall covered area and therefore the additional nodes can be shut down. If the threshold is chosen too large, too many nodes will be shut down, which will result in many isolated nodes and therefore a reduction of the covered area. If, however, the threshold is chosen too small, only few nodes will be shut down and the achievable enhancement of the overall lifetime will be only marginal. Therefore the threshold d_{Th} needs to be carefully chosen. The reactivation of the nodes is also idealized in this first approach by a very simple method: If a node runs out of battery, all neighboring nodes within the threshold range $(d_r \leq d_{Th})$ which were previously shut down are reactivated (Fig. 5.1c).

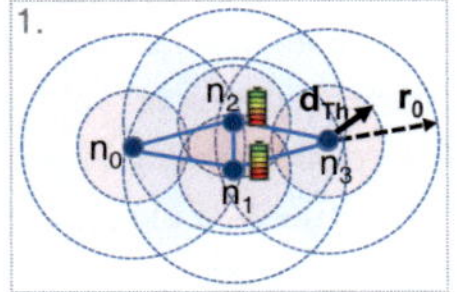
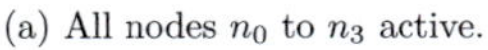
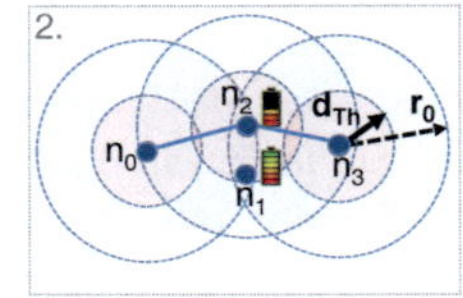
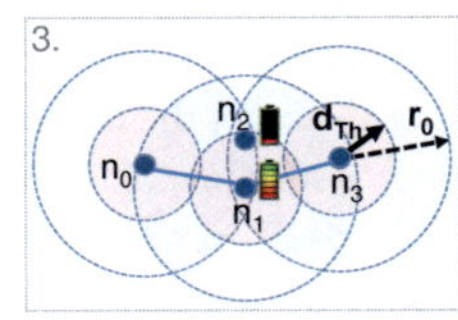

(a) All nodes n_0 to n_3 active.

(b) Node n_1 shut down into power saving mode. Data routed via n_2.

(c) n_2 out of power, n_1 reactivated. Data routed via n_1.

Figure 5.1: Sketch of the Lifetime Enhancement Algorithm (LEA) based on a distance threshold shut down mechanism.

Again the simulation parameters shown in Tab. 4.11 are used. The scenario with battery powered "MESH Routers" will be evaluated, where every MSTA has a battery power lifetime in between four to six hours in the simulation. The battery lifetime of the individual MSTAs again is calculated by using a Gaussian distribution with mean value of 5 hours and standard deviation of 0.5 hours. Simulations have been carried out with 102 MSTAs which leads according to Section 4.2.3 to an almost surely connected network in the considered area of $A = 250\,\text{m} \cdot 250\,\text{m}$ when using the Euclidean distance metric.

The flow chart in Fig. 5.2a details the algorithm that is used for this proof of concept study. First the coordinates of the Mesh Stations (MSTAs) are generated which are uniformly distributed in the area A. Isolated MSTAs will be shut down during the initialization as they will remain isolated during the simulation and do not enhance the coverage of the connected cluster. In the first part of the algorithm nodes are shut down according to the threshold condition in a sequential order. If the distance d_r of the currently considered MSTA to at least one of its neighboring MSTAs is less or equal than the threshold distance d_{Th} the current MSTA is shut down. If non of the other MSTAs is in range of the threshold distance to the current MSTA, the MSTA remains active. This is repeated for every MSTA. The second part of the algorithm emulates the lifetime of the system. If the battery lifetime (BLT) has expired of one of the remaining active MSTAs, this MSTA is shut down and all MSTAs within range less or equal than the threshold distance d_{Th} to the MSTA which has run out of battery are reactivated from sleep mode. The changing size of the covered area of the largest connected cluster will repeatedly be calculated over time which is not shown in Fig. 5.2a. The covered area itself is calculated using the Euclidean distance metric and using the unit disk connection model. This is repeated until all MSTAs are out of power. The whole procedure is repeated 1000 times following the approach of a Monte Carlo simulation.

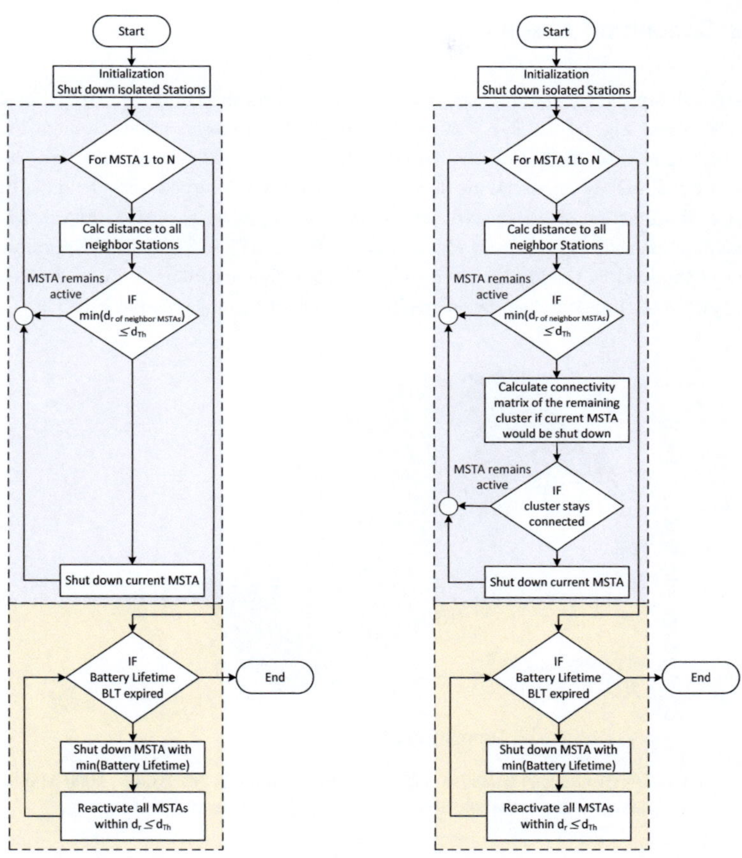

(a) Proof of concept Lifetime Enhancement Algorithm based on distance threshold shut down.

(b) Idealized version of the algorithm.

Figure 5.2: Flow chart of the proof of concept Lifetime Enhancement Algorithm (LEA) based on distance threshold shut down compared to an idealized approach.

In order to evaluate the performance of this algorithm the results will be compared to an idealized approach which is shown in the flow chart of Fig. 5.2b. In this approach MSTAs are shut down only, if the remaining set of MSTAs stays connected and no isolated MSTAs occur due to the shut down of the corresponding MSTA. This idealized approach assumes a central control unit which prevents MSTAs to shut down if isolated MSTAs or sub-clusters would arise and therefore is not a practical solution for the DRS, but is meant as a benchmark for the LEA only.

5.1.2 Simulation Results

Figure 5.3 shows the simulation result for a shut down threshold of $d_{Th} = 10\,$m, Fig. 5.4 for $d_{Th} = 20\,$m and Fig. 5.5 for $d_{Th} = 30\,$m. Figure 5.6 shows cross sections through Fig. 4.19 which corresponds to the initial lifetime estimation without sleep-wake algorithm shown in Section 4.3.1 as well as cross sections through Figs. 5.3, 5.4, 5.5 respectively, detailing the covered area as a function of the elapsed time in hours for $P_{\text{covered area}} = 90\%$. Figure 5.7 shows the cross sections corresponding to $P_{\text{covered area}} = 99\%$. In Figs. 5.6 and 5.7 the results that have been obtained by the idealized version of the algorithm according to Fig. 5.2b are shown in comparison to those obtained by the realistic version of the algorithm in Fig. 5.2a.

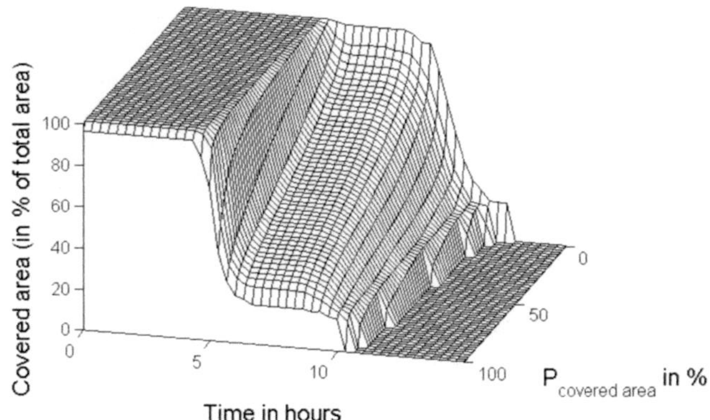

Figure 5.3: Covered area for $A = 250\,$m $\cdot\,250\,$m, radius $r_0 = 60\,$m, 102 Nodes, 1000 Monte Carlo simulation runs, threshold $d_{Th} = 10\,$m and Euclidean distance metric.

The results show that in case of a low threshold ($d_{Th} = 10\,$m) an extension of the lifetime is only given for small probabilities, i.e. $P_{\text{covered area}} < 5\%$ (see Fig. 5.3). The threshold is too small and only few nodes are shut down which is not sufficient to provide a significant extension of the lifetime by their later reactivation. For $d_{Th} = 20\,$m (Fig. 5.4) an extension of the lifetime to approximately 8.5 hours with probability $P_{\text{covered area}} = 90\%$ for a covered area of 90% is achieved. When using a threshold of $d_{Th} = 30\,$m, as shown in Fig. 5.5 an extension of the lifetime to approximately 9.5 hours with probability $P_{\text{covered area}} = 90\%$ for a covered area of 90% is achieved. Thus by an appropriate choice of the threshold d_{Th} lifetime of the network can almost be doubled.

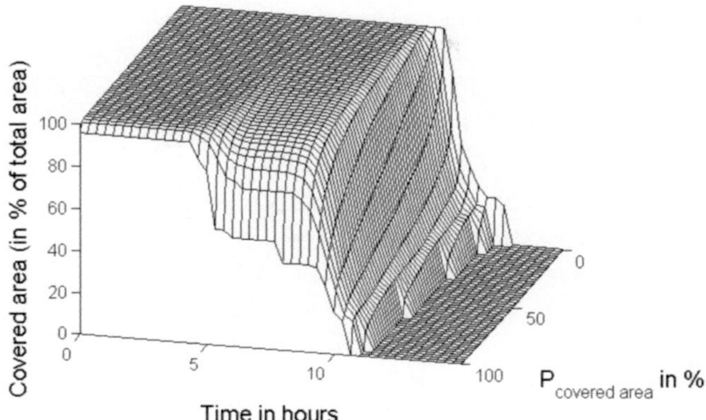

Figure 5.4: Covered area for $A = 250\,\mathrm{m} \cdot 250\,\mathrm{m}$, radius $r_0 = 60\,\mathrm{m}$, 102 Nodes, 1000 Monte Carlo simulation runs, threshold $d_{Th} = 20\,\mathrm{m}$ and Euclidean distance metric.

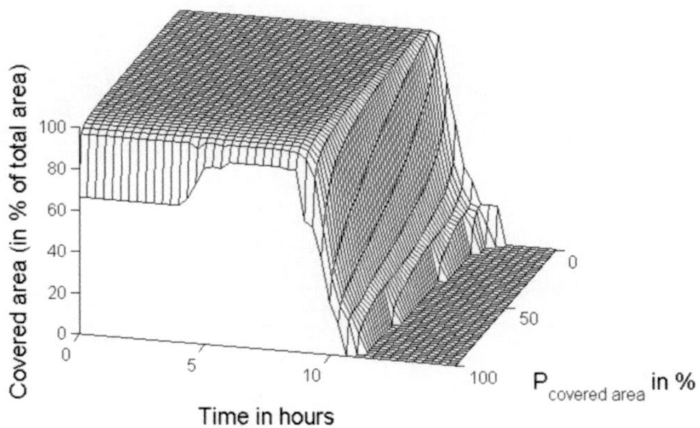

Figure 5.5: Covered area for $A = 250\,\mathrm{m} \cdot 250\,\mathrm{m}$, radius $r_0 = 60\,\mathrm{m}$, 102 Nodes, 1000 Monte Carlo simulation runs, threshold $d_{Th} = 30\,\mathrm{m}$ and Euclidean distance metric.

The flap at the beginning of time for $P_{\text{covered area}} > 99\%$ with $d_{Th} = 30\,\mathrm{m}$ (see Fig. 5.5) is due to an overly optimistic shut down of the nodes, which leads to a reduction of the covered area. A comparison with an ideal shut down of the nodes, i.e. nodes are only shut down if the connectivity remains unchanged and thus no isolated nodes arise (flow chart in Fig. 5.2b), confirms the results obtained when using the proposed algorithm (see Fig. 5.6), except that the flap at $d_{Th} = 30\,\mathrm{m}$ and $P_{\text{covered area}} > 99\%$ does not occur (see Fig. 5.7). This shows that isolated nodes only occur when the threshold distance d_{Th} is large. Therefore the threshold should not be chosen larger than $r_0/2$. The results confirm also that a central control unit is not needed when comparing the results to the ideal approach. Table 5.1 finally shows the detailed numbers resulting from the simulations for different time instances at $t = \{0, 300, 450, 510, 570, 600\}\,\text{min}$ and $P_{\text{covered area}} = \{90\%, 99\%\}$ when compared to the ideal approach.

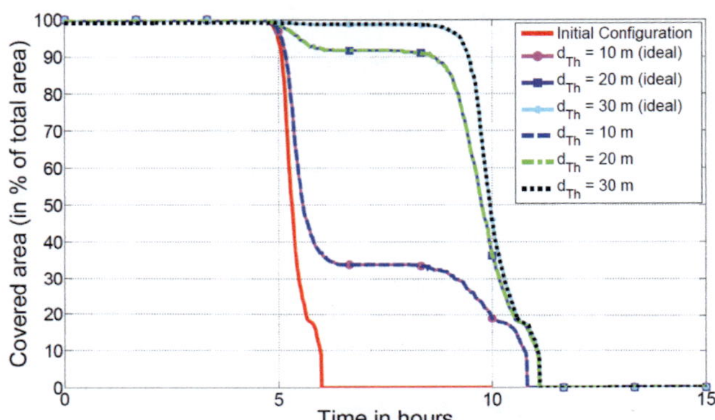

Figure 5.6: Cross sections for $P_{\text{covered area}} = 90\%$, covered area $A = 250\,\text{m} \cdot 250\,\text{m}$, radius $r_0 = 60\,\text{m}$, 102 Nodes, 1000 Monte Carlo simulation runs and Euclidean distance metric.

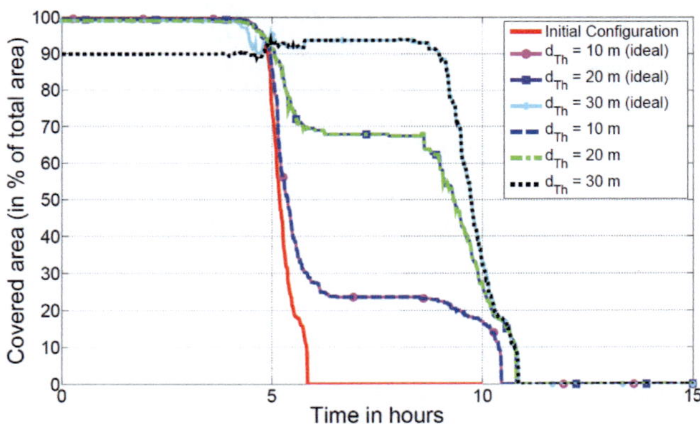

Figure 5.7: Cross sections for $P_{\text{covered area}} = 99\%$, covered area $A = 250\,\text{m} \cdot 250\,\text{m}$, radius $r_0 = 60\,\text{m}$, 102 Nodes, 1000 Monte Carlo simulation runs and Euclidean distance metric.

Table 5.1: Covered area (in percent of the total area) for $A = 250\,\text{m} \cdot 250\,\text{m}$, radius $r_0 = 60\,\text{m}$, 102 Nodes, 1000 Monte Carlo simulation runs, $P_{\text{covered area}} = \{90\%, 99\%\}$ for $t = \{0, 300, 450, 510, 570, 600\}\,\text{min}$ and Euclidean distance metric.

$P_{\text{covered area}}$	d_{Th}	0 min	300 min	450 min	510 min	570 min	600 min
Init. Config. 90%	–	100	95.78	0	0	0	0
90%	10 m	100	96.86	33.84	32.99	26.78	19.00
90% ideal	10 m	100	96.86	33.84	32.99	26.78	19.00
90%	20 m	99.99	98.75	91.69	90.61	66.16	36.28
90% ideal	20 m	100	98.75	91.69	90.61	66.16	36.28
90%	30 m	99.13	99.22	98.78	98.53	89.23	45.95
90% ideal	30 m	99.92	99.28	98.70	98.45	89.03	46.08
Init. Config. 99%	–	99.51	76.51	0	0	0	0
99%	10 m	99.49	85.72	23.53	23.53	19.61	16.93
99% ideal	10 m	99.51	85.72	23.53	23.53	19.61	16.93
99%	20 m	98.96	91.07	67.87	67.43	44.57	26.86
99% ideal	20 m	99.27	91.07	67.87	67.43	44.57	26.86
99%	30 m	89.92	93.63	93.72	93.07	67.81	32.22
99% ideal	30 m	99.04	95.25	93.72	93.07	67.67	32.49

5.1.3 Conclusion

The simulation-based proof of concept of the LEA-DRS algorithm showed that the lifetime of a DRS can be considerably extended depending on the node density while keeping the required coverage of the disaster area. The number of available nodes in the simulations was high enough to enhance the lifetime of the DRS. However, when a node ran out of power all previously shut down nodes within threshold range have been reactivated in this simplified approach. For a higher number of redundant nodes the lifetime enhancement thus is expected to be even higher which is, however, not part of the considerations in this section dealing with the proof of concept only. For the use case considered, lifetime could almost be doubled, which is key for bridging the time until professional rescuers arrive in the disaster region and will be able to bring up a dedicated network infrastructure. As key parameter for the lifetime enhancement algorithm a threshold beyond which devices individually will decide to switch off has been identified. The simulation-based proof of concept employs a rather high level of abstraction and will be further analyzed in the following sections by extending the algorithm into a distributed self organized sleep-wake algorithm. This algorithm works on basis of the IEEE 802.11s Mesh Peering Management (MPM) protocol by discovering an appropriate metric for the decision of the shut down and reactivation of nodes.

5.2 A Distributed Sleep-Wake-Algorithm for Lifetime Enhancement of Disaster Recovery Systems

In this section the first variant of the algorithm LEA-DRS shown in the previous section is extended by a mechanism allowing for a self organized shut down and power on of the individual nodes while still keeping coverage and connectivity of the network. The nodes in the proposed network in this solution act individually and a central server is not needed. The individual decision of the nodes is done without any knowledge of the whole network topology or the exact locations of the nodes. Thereby the decision if a node is needed can be based – following a first approach – on the distance to its neighboring nodes as shown in Section 5.1.

In Section 5.2.3 this algorithm is extended to a second version by an improved approach where the decision when a node changes its state from sleep to awake or vice-versa is based on the number of neighbors of the corresponding node, allowing for a much simpler implementation on the individual mesh nodes which can be based on the list of nodes to which a peer link is established and does not require any information on the RSSI.

5.2.1 Principle of the Distributed Algorithm

Figure 5.8 shows a simplified state chart of the algorithm acting at each node. In order to keep a compact notation the state chart corresponds to both versions of the LEA-DRS, to the one based on distance respectively on RSSI as well as to the one based on the number of neighbors according to the peer link list. All nodes are initialized with a random active time which is uniformly distributed. If the Station Active Timer (SAT) expires, the node will check whether the sleep condition is true or false. The sleep condition is false if either the number of neighbors of the corresponding node is less than or equal to the specified threshold (second version of the algorithm) or because the distance to all of its neighboring nodes is higher than the threshold condition (first version of the algorithm). Which condition will be used depends on the version of the LEA that is used. The node then remains in the Node Active state. If the sleep condition is true the node switches to the Node Sleep state.

If the Station Sleep Timer (SST) expires the node switches to the Node Active Update Peer Links (UPL) state. In this state the node activates itself and updates its peer link list. After the Peer Link update Timer (PLT) expires the node decides – based on the threshold condition – if it needs to stay active or could shut down itself again. After each change of state the timers (SAT or SST) are updated by a random timer value taken out of a uniform distribution of the specified time intervals.

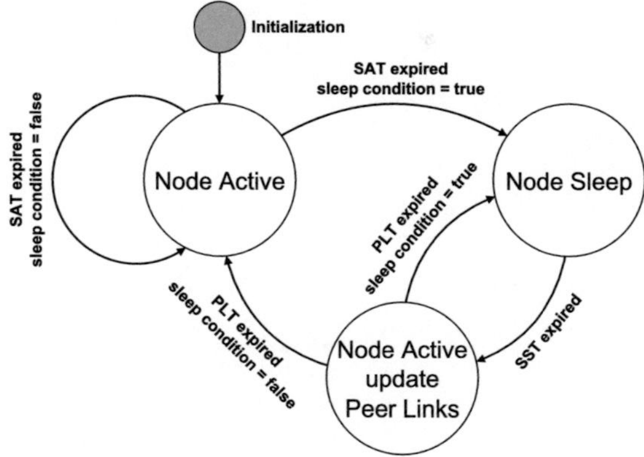

Figure 5.8: State chart of the LEA-DRS distributed sleep-wake algorithm.

5.2.2 Distributed Sleep-Wake Algorithm with Distance Metric

5.2.2.1 Methodology

Figure 5.9 shows the flow chart of the algorithm (executed in Matlab) which details the general state chart in Fig. 5.8. For each run of the Monte Carlo simulation all nodes are randomly placed in the area during initialization by using a discrete uniform distribution for the calculation of the coordinates of each node. The number of simulation runs was set to 1.000 with 100 MSTAs[1] each. Again the unit disk connection model is used, i.e. it is assumed that each node has connection to each of its neighbors if the distance to the neighboring node is less than or equal to $r_0 = 60\,$m. Isolated nodes are deactivated during initialization. The size of the area $A = 250\,\text{m} \cdot 250\,\text{m}$ has been chosen to limit the run time in the simulations executed in Matlab. In order to be able to scale the results to larger areas additionally the toroidal distance metric is used to eliminate the border effects. The algorithm is executed until all MSTAs are out of battery. The battery lifetime of the individual MSTAs is calculated by using the same distribution as before, i.e. a Gaussian distribution ($N(5, 0.5^2)$ hours) with mean value of 5 hours and standard deviation 0.5 hours. If the battery lifetime (BLT) of a MSTA has expired the MSTA is shut down and removed from the simulation. The new covered area of the remaining MSTAs is calculated after each change of state. The battery lifetime is running only when the nodes are in the Active state. The initial random active time was set to uniform $(5, 25)$ minutes. These values have been found by a heuristic approach which might not cover the optimum. The minimum active time of 5 min allows for establishing all Peer Links of all MSTAs before first

---

[1]In Chapter 6 network simulations will be carried out with positions of MSTAs as simulated in this section with 100 MSTAs and two additional user MSTAs leading to 102 MSTAs in total.

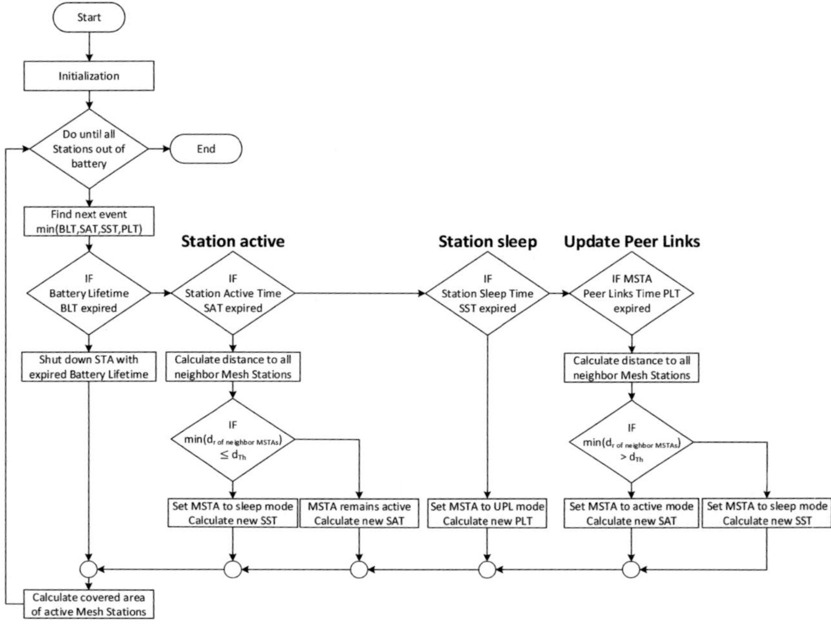

Figure 5.9: Flow chart (of the Matlab implementation) of the LEA distributed sleep-wake algorithm based on the distance metric.

MSTAs are evaluating the sleep condition and going to eventually switch the state. The interval of 20 min for the initial random active time allows for a certain degree of randomness, in order to avoid that all MSTAs are going to be switched off at the same time. After a change to the states Node Active or Node Sleep the timers SAT or SST are reset by a random time of the interval uniform $(8, 24)$ minutes which allows for a fast response of the algorithm when nodes are running out of power. The Peer Link update Timer has a fixed value of 15 sec which is long enough to ensure that all peer links would have been established, to all neighboring nodes during that time, in view of the IEEE 802.11s protocol. It should be noted that the IEEE 802.11s Mesh Peering Management protocol is not part of this Matlab implementation. Additionally, it is assumed that the distance d_r could be estimated by the Received Signal Strength Indicator (RSSI) of the messages exchanged during the peer link establishment. A node which is in the Node Active Update Peer Links (UPL) state is not included in the calculations of the covered area even though it is active at the specific time. This avoids overly optimistic results by neglecting MSTAs which are shut down again after expiration of the Peer Link Update Timer. The covered area is calculated by considering all active nodes which form the largest cluster[2] of

---

[2]By largest cluster again the cluster covering the largest contiguous area is meant.

connected nodes. Isolated nodes and smaller clusters are neglected in the simulation analysis. An overview of the simulation parameters is given in Tab. 5.2.

Table 5.2: Simulation parameters of distributed sleep-wake algorithm with distance metric.

Simulation Environment	
Area A	$250\,\mathrm{m} \cdot 250\,\mathrm{m}$
Connection Model	Unit Disk model with Euclidean or toroidal distance metric
Transmission Range r_0	$60\,\mathrm{m}$
Number of MSTAs	max. 100 active MSTAs
Number of Monte Carlo runs	1000
Battery Lifetime (BLT)	BLT_i = random variable with probability distribution $N(5, 0.5^2)$ hours
LEA	
Initial active time	uniform $(5, 25)\,\mathrm{min}$
Station Active Timer (SAT)	uniform $(8, 24)\,\mathrm{min}$
Station Sleep Timer (SST)	uniform $(8, 24)\,\mathrm{min}$
Peer Link update Timer (PLT)	$15\,\mathrm{sec}$

5.2.2.2 Simulation Results

Figure 5.10 shows the results with shut down thresholds of $d_{Th} = 10\,\mathrm{m}$, $d_{Th} = 20\,\mathrm{m}$ and $d_{Th} = 30\,\mathrm{m}$ with Euclidean distance metric (edm, solid lines), compared to the initial configuration without sleep-wake algorithm, detailing the covered area as a function of the elapsed time in hours for $P_{\text{covered area}} = 90\%$. Additionally, the results are compared to the equivalent results when using the toroidal distance metric (tdm, dashed lines).

The results show that with a threshold of $d_{Th} = 10\,\mathrm{m}$ (red solid line) the covered area is almost 100% for the first 5 hours, however, the covered area is strongly reduced after the first MSTAs run out of battery at around 5 hours. The overall lifetime is not significantly extended at a threshold of $10\,\mathrm{m}$ compared to the initial simulation without sleep-wake algorithm even when using the toroidal distance metric (red dashed line). The threshold of $d_{Th} = 10\,\mathrm{m}$ is too small, as only few nodes are shut down by the algorithm and therefore only few nodes are available for reactivation after the first nodes run out of power. With a threshold of $d_{Th} = 20\,\mathrm{m}$ (blue solid line) the covered area is more than 95% for 90% of the simulations for the first 5 hours. However, after the first nodes run out of battery the covered area is reduced to about 80%. This also holds for the simulation with threshold of $d_{Th} = 30\,\mathrm{m}$ (green solid line). The reduction in coverage to below 80% at about 5 hours essentially is a result of the Station Sleep Timer (SST) which shows the response time of the algorithm when nodes are out of battery and nodes previously shut down have to be reactivated.

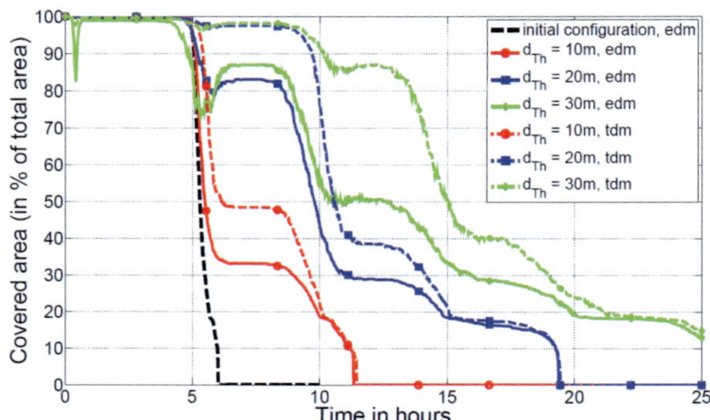

Figure 5.10: LEA-DRS distributed sleep-wake algorithm: Covered area for $A = 250\,\text{m} \cdot 250\,\text{m}$, radius $r_0 = 60\,\text{m}$, 100 Nodes, 1000 Monte Carlo simulation runs, comparing various thresholds of the distance for Euclidean and toroidal distance metric at $P_{\text{covered area}} = 90\%$.

The dip at the beginning of the simulation which is most significant for $d_{Th} = 30\,\text{m}$ indicates the settling process of the algorithm which is dependent on the initial active time and the time interval of the state timers (SAT and SST). Short timer values lead to a fast settling of the algorithm, however, with the drawback of many on/off changes of the nodes at the beginning and also during run time. Higher values lead to less management traffic overhead and thus less network load. However, if a node runs out of battery the response time for a node which has to take the place of the previously failed node is also increased. As a possible solution, an exponential increase of the timer values to a maximum value can be considered to reduce the settling time at the beginning and to maintain an appropriate response time during operation.

By excluding border effects for $d_{Th} = 20\,\text{m}$ and $d_{Th} = 30\,\text{m}$ when using the toroidal distance metric an extension of the lifetime to about 10 hours is reached which is in the range of the results obtained in Section 5.1, however with Euclidean distance metric. The better results obtained by the proof of concept in Section 5.1 can be explained by the idealized approach used for the proof of concept within which all previously shut down nodes are reactivated within the range of a failed node at the same time. On the other hand, when using the distributed sleep-wake algorithm a covered area of still 50% after 15 hours can be achieved ($d_{Th} = 30\,\text{m}$, tdm) which could not be reached in the proof of concept simulation.

5.2.3 Distributed Sleep-Wake Algorithm with Metric Based on the Minimum Number of Neighbors

This section focusses on implementability of the distributed sleep-wake algorithm by the second version of the algorithm, following an alternative approach where the number of neighbors (NoN) of a MSTA are investigated rather than the distance which has been evaluated for the first version of the algorithm in the previous sections. The number of neighbor MSTAs can be obtained with the help of the Mesh Peering Management (MPM) protocol (see Section 2.2.3). Each MSTA to which a peer link successfully is established is a neighbor MSTA of the corresponding MSTA. Thus the number of neighboring MSTAs simply can be obtained by the peer link list of each MSTA. In contrast to the approach with distance metric only MSTAs to which a peer link successfully has been established are taken into account. Additionally, there is no need to estimate the distance to neighboring MSTAs, e. g. through RSSI considerations.

5.2.3.1 Methodology

Figure 5.11 shows the flow chart of the second version of the algorithm. In contrast to the flow chart shown in Fig. 5.9, now the sleep condition is evaluated by considering the number of neighbors of the corresponding MSTA. If the actual MSTA is active and the number of neighbors is above the specified threshold value NoN_{Th}, the actual MSTA is set to the sleep mode. Otherwise it remains active. The initial random active time was again set to uniform $(5, 25)$ min. The simulation parameters are the same as shown in Tab. 5.2 with the number of Monte Carlo simulation runs, namely 1.000 each with 100 MSTAs. The simulations are carried out for three different thresholds for the number of neighbors namely $\mathrm{NoN}_{Th} = 6, 7$ and 8, the values of which are inspired by the results for the so called "magic numbers" discussed in Section 3.1.3.

5.2.3.2 Simulation Results

Figure 5.12 shows the covered area as a function of the elapsed time in hours and the probability $P_{\text{covered area}}$ for a shut down threshold of $\mathrm{NoN}_{Th} = 7$ with Euclidean distance metric and Fig. 5.13 for $\mathrm{NoN}_{Th} = 7$ with toroidal distance metric.

Figure 5.14 shows cross sections through Figs. 5.12 and 5.13 as well as additional results with shut down thresholds of $\mathrm{NoN}_{Th} = 6$ and 8 compared to the initial configuration without sleep-wake algorithm, detailing the covered area as a function of the elapsed time in hours for $P_{\text{covered area}} = 90\%$.

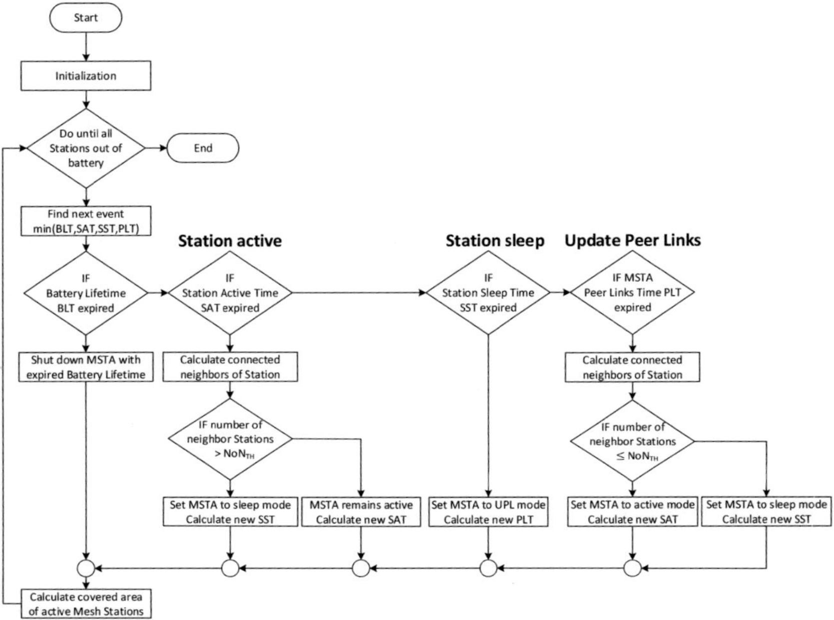

Figure 5.11: Flow chart (of the Matlab implementation) of the LEA distributed sleep-wake algorithm with metric derived from the number of neighbors (NoN).

The results show that with a threshold of 6 neighbors (red solid line) the covered area is only slightly more than 80% for the first 5 hours and the covered area is further reduced after the first MSTAs run out of battery at around 5 hours. With a threshold of 7 (blue solid line) and 8 (green solid line) the covered area is more than 95% for 90% of the simulations for the first 5 hours. The threshold of 6 thus turns out to be too stringent as too many nodes are shut down during network start-up resulting in a lower coverage. However, excluding border effects by employing the toroidal distance metric (dashed lines) all variants analyzed in Fig. 5.14 show a covered area of more than 90% for almost 10 hours.

The considerable differences between the simulation results for the Euclidean and for the toroidal distance metric are readily explained by the fact that the simulated area with 250 m · 250 m is rather small and nodes at the edges of the area have less neighbor nodes than in the middle. This leads to more nodes in the on state at the edges of the area. If the first nodes run out of battery less nodes are available at the edges and the covered area shrinks to the center of the area. Nodes around the center of the network, however, will still have connection to each other. This can be seen in Fig. 5.15 when comparing the positions of the active nodes of all simulation runs at time $t = 600$ min with Euclidean and toroidal distance metric.

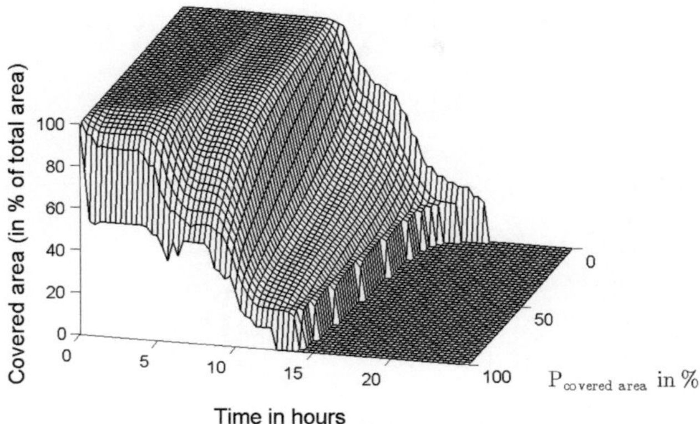

Figure 5.12: LEA-DRS distributed sleep-wake algorithm: Covered area for $A = 250\,\text{m} \cdot 250\,\text{m}$, radius $r_0 = 60\,\text{m}$, 100 Nodes, 1000 Monte Carlo simulation runs, threshold $\text{NoN}_{Th} = 7$, Euclidean distance metric.

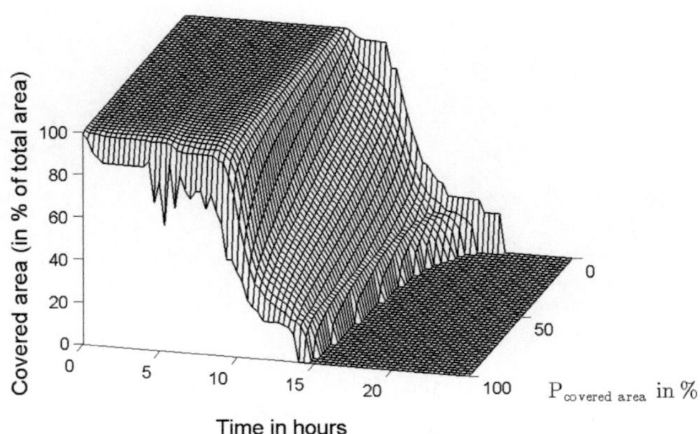

Figure 5.13: LEA-DRS distributed sleep-wake algorithm: Covered area for $A = 250\,\text{m} \cdot 250\,\text{m}$, radius $r_0 = 60\,\text{m}$, 100 Nodes, 1000 Monte Carlo simulation runs, threshold $\text{NoN}_{Th} = 7$, toroidal distance metric.

It can be seen in Fig. 5.15a from the nodes active after $t = 600\,\text{min}$ that the covered area is shrinked to the center of the total area. However, when using the toroidal distance metric as shown in Fig. 5.15b still reasonable results can be achieved which are scalable to larger areas.

It has been shown that with a threshold of 7 and 8 for the number of neighbors a sufficient network coverage can be achieved. For the use case considered in this work, lifetime could almost

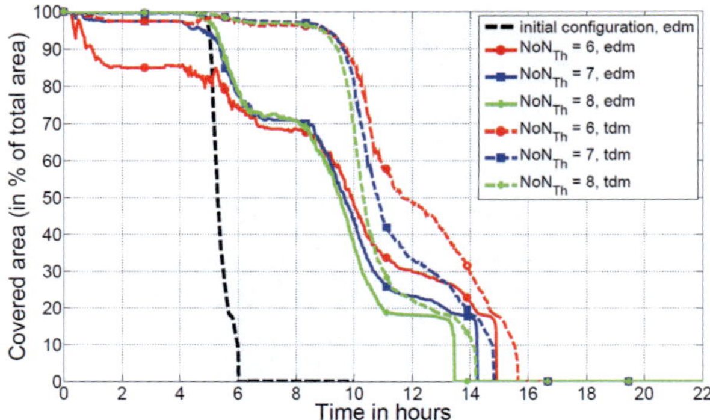

Figure 5.14: LEA-DRS distributed sleep-wake algorithm: Covered area for $A = 250\,\text{m} \cdot 250\,\text{m}$, radius $r_0 = 60\,\text{m}$, 100 Nodes, 1000 simulation runs, comparing various thresholds for the number of neighbors with Euclidean and toroidal distance metric at $P_{\text{covered area}} = 90\%$.

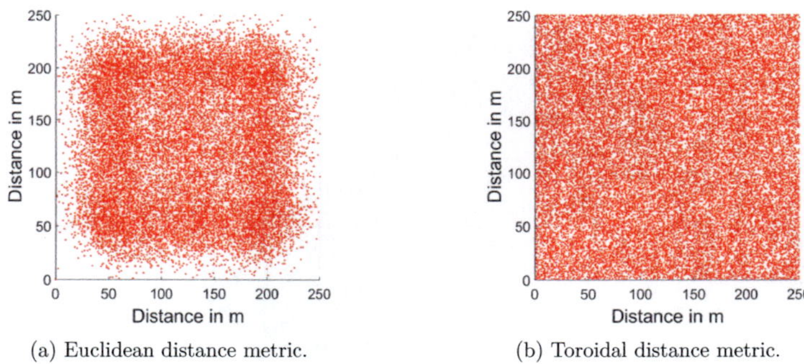

(a) Euclidean distance metric.
(b) Toroidal distance metric.

Figure 5.15: Cumulative representation of the active nodes for 1000 Monte Carlo simulation runs at time $t = 600\,\text{min}$ in the area $A = 250\,\text{m} \cdot 250\,\text{m}$, with 100 initial nodes per simulation run, with Euclidean and toroidal distance metric.

be doubled. Again it should be noted, that the number of stations has been restricted to 100, which is by far less than what can be expected from measurements of available stations in real environments. From [17] about 300 stations can be expected within an area of $250\,\text{m} \cdot 250\,\text{m}$ which would lead to an expected lifetime of the DRS of approximately 30 hours at high coverage and complete shut down of the DRS only after approximately 45 hours. Assuming increasing battery capacity in the future, covering the first golden 72 hours even with user equipment based DRS seems rather reasonable.

Figure 5.16 shows the mean number of neighbors for the active nodes which result from the distribution of nodes as a result of the execution of the algorithm for the toroidal and Euclidean distance metric.

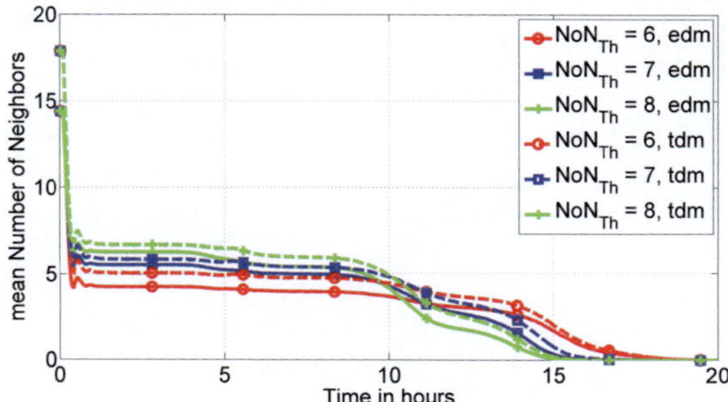

Figure 5.16: LEA-DRS distributed sleep-wake algorithm: Mean number of neighbors of each individual node for Euclidean and toroidal distance metric for $A = 250\,\text{m} \cdot 250\,\text{m}$, radius $r_0 = 60\,\text{m}$, 100 Nodes, 1000 Monte Carlo simulation runs, comparing various thresholds for the number of neighbors.

At the beginning of the simulation the mean number of neighbor nodes is about 18 for the toroidal and 14 for the Euclidean distance metric. After the algorithm has settled the actual mean number of neighbor nodes (dashed lines) when using the toroidal distance metric, excluding border effects, is about one node below the NoN threshold which is a result of the on-off behavior of the algorithm. This means that more nodes than should be – according to the corresponding threshold – are in the sleep mode which is also an indication for a slight ping-pong effect. When using the Euclidean distance metric the actual mean number of neighbor nodes (solid lines) is about two nodes below the threshold which is also a result of the on-off behavior of the algorithm and additionally due to the border effect as nodes at the edges of the area have less neighbor nodes than in the middle and some isolated nodes may arise. In Section 4.2.2 it has been shown that the mean number of neighbor nodes provides a good metric for the scalability of the network to larger areas. Therefore, when using the NoN metric for the distributed sleep-wake algorithm the obtained results are expected to show reasonable scalability to larger areas as the number of active nodes is only based on this metric.

Figure 5.17 shows the mean number of active MSTAs for all configurations over the complete simulation time. The mean number of active MSTAs for the initial configuration (exact values are shown in Tab. 6.3) is slightly below the maximum of 100 MSTAs. The reason is that isolated nodes are deactivated before the start of the simulation and therefore not taken into account. When using the Euclidean norm, the number of active MSTAs in the first phase is higher, since fewer MSTAs can be switched off at the borders of the area. Thus the available number of active MSTAs in the second phase for $t > 5\,\text{h}$ is lower compared to the results when using the toroidal distance metric.

The results of active MSTAs are also in a range which would have been expected in view of the results shown in Tab. 4.5.

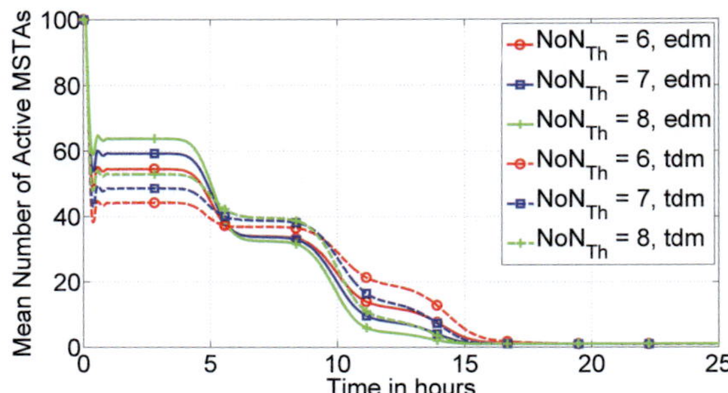

Figure 5.17: Mean number of active MSTAs according to the Matlab implementation of the LEA distributed sleep-wake algorithm with number of neighbors (NoN) metric and $\text{NoN}_{Th} = 6,7,8$.

Figure 5.18 further shows the comparison with an ideal approach of the proposed algorithm. To this end the algorithm shown in the flow chart of Fig. 5.11 is extended in order to make sure, that a MSTA in the Station active or Update Peer Links state is only set to sleep if the remaining MSTAs leave unchanged the biggest cluster of MSTAs in order to avoid isolated MSTAs or smaller subclusters. This idealized approach assumes an superordinate instance controlling the MSTAs.

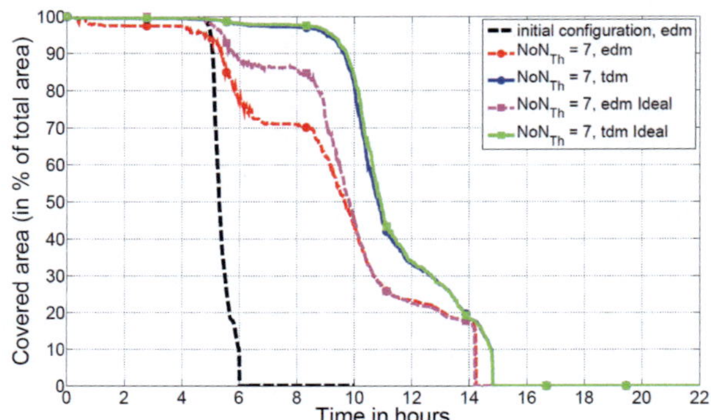

Figure 5.18: LEA-DRS distributed sleep-wake algorithm: Covered area for $A = 250\,\text{m} \cdot 250\,\text{m}$, radius $r_0 = 60\,\text{m}$, 100 Nodes, 1000 Monte Carlo simulation runs, comparison with ideal approach for Euclidean and toroidal distance metric at $P_{\text{covered area}} = 90\%$.

This idealization doesn't lead to any differences as long as a threshold of 7 neighboring nodes and the toroidal distance metric are considered. By including border effects through the Euclidean distance metric and due to the relatively small simulation area of $250\,\text{m} \cdot 250\,\text{m}$ some differences can be seen during runtime.

The better results with the idealized approach indicate that in the simulation with Euclidean distance metric some isolated MSTAs appear. However, these isolated MSTAs may mainly occur at the boundary of the area while the coverage at the center remains unchanged. The results show by excluding border effects that the simulation with a threshold of 7 neighbor nodes approaches the ideal case. This confirms the conclusion that a central control unit is not needed.

5.2.4 Conclusion

The results obtained by the LEA-DRS sleep-wake algorithm with number of neighbors threshold show comparable results to those shown in Section 5.2.2 with distance threshold. In view of implementability the algorithm with number of neighbors threshold is preferred because the number of neighbors of each MSTA is a parameter known to each MSTA through the peer link list according to IEEE 802.11s and allows a much simpler implementation. RSSI on the other hand is a parameter which is dependent on the manufacturers implementation because there are no uniform standards. Additionally, the value my change rapidly due to fading effects which are not considered in the previous simulations. In contrast a MSTA is only listed in the peer link list if a link could have been successfully established by the two way handshake mechanism (see Section 2.2.3) which successfully has to be carried out even under fading conditions. Of course a MSTA which is shut down or has left the mesh network needs to be removed from the peer link list in order to avoid the shut down of MSTA based on a wrong number of neighbor MSTAs. This will be accomplished by the peer link close process as shown in the finite state machine of the MPM protocol in [35, Fig. 14-2]. The influence of the network protocol an the performance of the DRS will be further discussed in Chapter 6.

Part II

System Implementation and Validation

Chapter 6

Network Performance Analysis of the Lifetime Enhancement Algorithm (LEA)

Contents

This chapter validates that with the reduced set of nodes which results from the proposed distributed sleep-wake algorithm introduced in Chapter 5 communication in the network is still possible at sufficient performance. The network simulator ns-3 as described in Section 3.2 is used to validate the results from Chapter 5 by evaluating a VoIP connection of two individual rescuers including the whole protocol stack. The IEEE 802.11s mesh model which is used here has been evaluated in Section 3.2.2. In order to compare the results with those shown in Chapter 5, a fixed transmission radius of $r_0 = 60\,$m has been chosen using the ns-3 Range-Propagation-Loss-Model. The WMN topology which is based on IEEE 802.11s is set up with the same coordinates of the MSTAs as simulated in Matlab with usual Euclidean distance metric. Two additional nodes which represent the user nodes are randomly placed in the topology as communication partners for the validation of a communication link. The simulations have been carried out with 1000 Monte Carlo simulation runs for each configuration of the sleep-wake algorithm with threshold for the number of neighbors $NoN_{Th} = 6$, 7 or 8 as well as for the initial configuration without sleep-wake algorithm and thus the full set of activated nodes.

First the network performance will be analyzed in a static scenario at specific instants in time with the reduced set of nodes obtained from the Matlab simulations. These simulations will be called static network simulations as no changes to the number and positions of the nodes in the network do occur. Second, the dynamic network performance of the distributed sleep-wake algorithm will be analyzed in the network simulation which also includes additional network traffic (e. g. peer link establishment and routing) of the IEEE 802.11s MAC due to the activation and deactivation of MSTAs. These network simulations will be called dynamic network simulations as the number of active nodes will change due to the execution of the sleep-wake algorithm in the network simulator itself. Section 6.3 further compares the influence of different QoS access categories of IEEE 802.11 as described in Section 2.2 and Appendix A.

6.1 Network Performance Analysis for Fixed Topologies Resulting from the Lifetime-Enhancement Sleep-Wake Algorithm

6.1.1 Static Network Simulation Setup

Assumed is a VoIP communication link between the two user nodes. As there is no VoIP communication model in ns-3, the communication link is modeled by two individual UDP data streams. Assuming a G.711 VoIP codec with default packetization rate (50 pps) a new VoIP packet is generated every 20 ms. A set of protocols which are designed to deliver communication services over the IP network is needed. These are signalling protocols, e. g. the Session Initiation Protocol (SIP) and the RTP protocol for the audio transport. RTP is dedicated to convert analog voice signals into data packets which are encapsulated in UDP packets. The data transport is augmented by a control protocol (RTCP) to allow monitoring of the data delivery [118, RFC 3550]. The payload of each VoIP packet is 160 bytes. With the IP, UDP and RTP headers $(20 + 8 + 12$ bytes) included, this packet becomes 200 bytes in length. There is currently no RTP protocol in ns-3 but the corresponding header can be virtually added as additional payload such that the total payload is set to 172 bytes in the simulation. By taking into account the IEEE 802.11 QoS Data and Logical-Link Control header the packet becomes

250 bytes in total (see Tab. 6.1). For the simulation of voice calls links with fixed data-rate are considered regardless of the amount of the conversation which is speech and how much is silence. The influence of higher layer protocols which are for example used to establish connections over the network as the Session Initiation Protocol are not considered. Also ideal establishment of the communication link is assumed. If not stated otherwise, the default QoS access category best effort (AC_BE) is used in the simulations.

Table 6.1: Packet compilation of VoIP packet in ns-3.

Protocol	Length (bytes)
IEEE 802.11 QoS Data	42
Logical-Link Control	8
IPv4	20
UDP	8
Payload with emulated RTP header	172
Packetsize	250

Table 6.2: Simulation parameters in ns-3 for the static network simulations.

Simulation Environment	
Area A	$250\,\mathrm{m} \cdot 250\,\mathrm{m}$
Propagation Model	RangePropagationLossModel
Max Range	$60\,\mathrm{m}$
Number of MSTA	2 User MSTA, max 100 active MSTAs
IEEE 802.11s Peer Link	
MaxBeaconLoss	20
MaxRetries	4
MaxPacketFailure	25
MaxNumberOfPeerLinks	32
IEEE 802.11s HWMP Protocol	
Dot11MeshHWMPmaxPREQretries	5
UnicastPreqThreshold	{1,10}
UnicastDataThreshold	{1,10}
Communication Link	
Tx rate	$80\,\mathrm{kbps}$ ($68.8\,\mathrm{kbps}$ payload)
Packetization Interval	$20\,\mathrm{ms}$
Voice Payload in Bytes	172
Start Time	uniform $[5,35]$ sec
Stop Time	uniform $[155,185]$ sec
Transport Protocol	UDP

The parameters of the ns-3 simulation are shown in Tab. 6.2. The area $A = 250\,\mathrm{m} \cdot 250\,\mathrm{m}$ is specified as shown in the coverage simulations in Section 5.2.3 with two additional user

MSTAs as communication partners and a maximum of 100 relay MSTAs. The VoIP call starts randomly for each simulation run within a time interval from 5 to 35 sec after the beginning of the simulation and has a random duration of 120 up to 180 sec. The second UDP data stream starts randomly in between 100 ms to 200 ms after the first data stream. The WMN operates on a single frequency channel. At the beginning of the simulation each MSTA is setting up a peer link to all of its neighboring MSTAs according to the peer link management protocol. After the establishment of all peer links, which is realized within a few seconds after the beginning of the simulation, the DRS network is assumed to be established as well. It should be noted, that in a dense network some of the nodes might interfere with each other and not all peer links would be established resulting in a less connected network.

The network simulator ns-3 further allows the setting of the UnicastPreqThreshold and Unicast-DataThreshold. The UnicastPreqThreshold allows the adjustment of the maximum number of PREQ receivers (neighbor MSTAs) when path requests are sent as a chain of unicast messages. The UnicastDataThreshold refers to the maximum number of broadcast receivers, when a broadcast message is sent as a chain of unicasts. An example where this parameter has an effect is the Address Resolution Protocol (ARP). Both values may have significant influence on the overall performance of the network. If a broadcast message is sent as unicast, the unicast message is acknowledged by the receiver and reception is ensured. However, if the number of neighbor nodes is high in a dense network this comes with the drawback of a significant increase in management traffic overhead and therefore an appropriate parametrization is needed. In the network simulation two thresholds {1,10} are chosen in order to compare if broadcast messages which are sent as unicast or sent as intended have an influence on the network performance. For the simulation with all broadcast messages sent as unicast a threshold value of 10 has been chosen meaning that messages from a MSTA with less or equal than 10 neighbor MSTAs are sent as unicast. If an acknowledge is not received following a sent unicast PREQ the PREQ will be repeated four times according to the specified ns-3 parameter MaxRetries.

As there is no root node in the WMN topology of a disaster area, the reactive on-demand mode of the hybrid wireless mesh protocol HWMP is used in order to establish a route between the two communicating nodes.

The static ns-3 simulation is set up with the coordinates of the MSTAs as simulated in the coverage simulation in Matlab using the Euclidean distance metric (Section 5.2.3) at a low number of specific time instances, where only nodes which are in the Node Active state at these specified time instances T_0, T_1, T_2, T_3 and T_4 are used and remain active during the network simulation. The simulations are carried out for the initial set up at time T_0=0 min with all nodes active and when using the sleep-wake algorithm at times T_1=22 min, T_2=80 min, T_3=300 min and T_4=600 min. The covered area for various thresholds for the number of neighbors at $P_{\text{covered area}} = 90\%$ has been shown in Fig. 5.14 of Section 5.2.3. The specific time instances for the network performance simulation in ns-3 are chosen as follows: At time T_1 the network is in the settling process according to the sleep-wake algorithm. At time T_2 the algorithm has settled. At time T_3 nodes previously active run out of power leading to a reduced overall coverage. At time T_4 the set of nodes is further reduced as most nodes are already out of power.

6.1.2 Static Network Simulation Results

The interpretation of the results is based on the QoS requirements of VoIP according to
[105]:

- packet loss should be no more than 1 percent,

- one-way latency (mouth to ear) should be no more than 150 ms,

- average one-way jitter should be targeted at less than 30 ms.

A successful call is assumed if the requirements for latency, jitter and packet loss are met for
both of the individual data links (Eq. (6-1)). A call is considered as dropped if at least one of
the three conditions is not fulfilled for at least one of the two links. Denoting the individual
conditions for a successful call on the two data links required in the simulation for modeling
the bi-directional VoIP call by (Successful_Call$_1$) and (Successful_Call$_2$) we can write the
condition for a successful VoIP call formally as:

$$\text{Successful\_Call}_n = (\text{Packet loss} \leq 0.01) \ \& \ (\text{mean Delay} \leq 150\,\text{ms}) \ \& \ (\text{mean Jitter} \leq 30\,\text{ms})$$

$$\text{Successful\_Call} = (\text{Successful\_Call}_1) \ \& \ (\text{Successful\_Call}_2) \tag{6-1}$$

Figure 6.1 shows the probability of a successful call for the static setup at times T_1, T_2, T_3 and
T_4. Additionally, Tab. 6.3 shows the exact values and the comparison with the initial setup at
time T_0 with all nodes active.

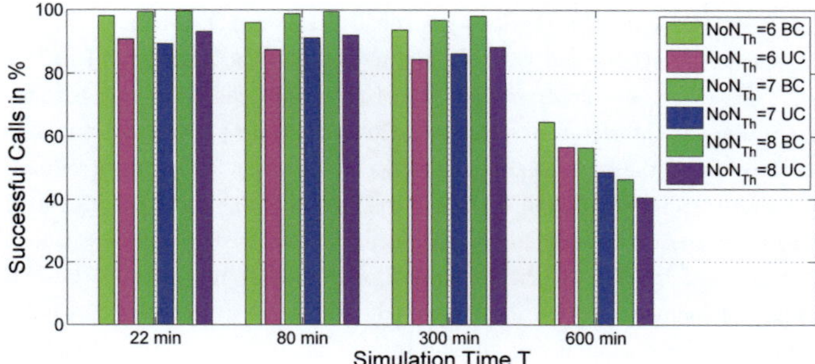

Figure 6.1: Probability of successful calls at T_1, T_2, T_3, T_4 and number of neighbors (NoN)
metric with NoN$_{Th}$ = 6,7,8 and comparison of the unicast (UC) (i. e. UnicastPreq-
Threshold and UnicastDataThreshold = 10) and broadcast (BC) (i. e. UnicastPreq-
Threshold and UnicastDataThreshold = 1) results.

At T_1 which is in the settling time of the algorithm the probability of a successful call is more than 98% for all configurations with UnicastPreqThreshold $= 1$. At T_2 the probability of a successful call is in between 96% to 99.5% with the lowest probability for $NoN_{Th} = 6$. At T_3 the probability drops to values between 93.7% and 98.1% and at T_4 the configuration with $NoN_{Th} = 8$ drops to the lowest value of about 46%.

Table 6.3: Probability of successful calls – static setup.

| | Ratio of Successful Calls | | mean number of neighbors | mean number of active MSTAs |
	PREQ broadcast	PREQ unicast		
NoN_{Th}=n.a., T_0=0 min	98.8%	96.7%	14.4	99.98
$NoN_{Th} = 6$, T_1=22 min	98.1%	90.8%	4.39	48.49
$NoN_{Th} = 7$, T_1=22 min	99.4%	89.4%	5.64	53.82
$NoN_{Th} = 8$, T_1=22 min	99.7%	93.2%	6.33	58.94
$NoN_{Th} = 6$, T_2=80 min	96.0%	87.5%	4.24	54.39
$NoN_{Th} = 7$, T_2=80 min	98.8%	91.2%	5.52	59.19
$NoN_{Th} = 8$, T_2=80 min	99.5%	92.1%	6.27	63.71
$NoN_{Th} = 6$, T_3=300 min	93.7%	84.5%	4.08	44.78
$NoN_{Th} = 7$, T_3=300 min	96.8%	86.3%	5.22	47.69
$NoN_{Th} = 8$, T_3=300 min	98.1%	88.3%	5.85	50.25
$NoN_{Th} = 6$, T_4=600 min	64.7%	56.7%	3.61	22.55
$NoN_{Th} = 7$, T_4=600 min	56.5%	48.7%	4.25	19.76
$NoN_{Th} = 8$, T_4=600 min	46.5%	40.6%	4.12	16.25

Figure 6.1 further compares the results with the UnicastPreqThreshold $= 10$ meaning that PREQs are sent as unicast instead as broadcast messages if the number of PREQ receivers is less than 10. The result is a lower probability of successful calls due to a higher amount of management traffic and interference in the scenarios with PREQ messages sent as unicast. For the use case considered the number of successful calls is lower for all observed scenarios and settings. Therefore the setting of the UnicastPreqThreshold and UnicastDataThreshold is not useful to be higher than 1 for the considered scenarios. It has to be noted that the setting of the transmission of PREQs via either unicast or broadcast was part of the IEEE 802.11s Draft and was not included in the final standard. The final standard allows only broadcast messages. The following simulations will be carried out with a UnicastPreqThreshold and UnicastDataThreshold equal to 1.

Table 6.3 further shows the actual value of the mean number of neighbors as originally shown in Fig. 5.16 and the mean number of active MSTAs. This can be compared to Fig. 5.17 in Section 5.2.3 which showed the mean number of active MSTAs for all configurations over the complete simulation time. As the nodes of the ns-3 simulation are taken as the MSTAs in the active state, the number in Tab. 6.3 are exactly the ones that are obtained from the Matlab simulations in Section 5.2.3. The mean number of active MSTAs for the initial configuration (first line in Tab. 6.3) is slightly below the maximum of 100 MSTAs. The reason is that isolated nodes are deactivated before the start of the simulation and therefore not taken into account.

6.2 Network Performance when Operating the Lifetime-Enhancement Sleep-Wake Algorithm under Network Conditions

6.2.1 Dynamic Network Simulation Setup

The dynamic simulation reflects the implementation of the distributed sleep-wake algorithm as described in Section 5.2.3 and thus will include all the network overhead required for peer link establishment as well as the influence of the network protocol, especially the peer link management protocol and HWMP, on the distributed sleep-wake algorithm.

First we need to clarify a few implementation issues of the distributed sleep-wake algorithm which cannot entirely be realized on basis of the IEEE 802.11s protocol.

In the Matlab implementation all nodes are in a dedicated state which is known to each of the nodes. According to IEEE 802.11s [35], if a MSTA leaves the mesh Basic Service Set (BSS) of which it is a member, it should explicitly close all of its active mesh peerings by using Mesh Peering Close frames. However, if a MSTA leaves abruptly the mesh BSS, e. g. because a mobile MSTA leaves physically the network or runs out of power, the MSTA has no time to set up mesh peering close frames. This situation is not explicitly handled in IEEE 802.11s. In this case a peer link would always be listed as established if outdated links are not deleted. According to the proposed sleep-wake algorithm, if the Station Active Timer (SAT) expires and a node verifies whether the sleep condition is true or false this decision could therefore be based on a wrong number of active peer links resulting in too many nodes to shut down. In principle the ns-3 mesh model implemented two ways of handling this. Links can be closed either due to a maximum beacon loss of successive beacons or due to a maximum packet loss of successive packets. If one of the parameters exceeds the threshold the link will be closed. The maximum beacon loss was set to 20. With a beacon interval of 1.024 sec a link will be closed after 20.48 sec.

Furthermore, nodes which are in the Node Active update Peer Links state could also lead to a wrong decision of nodes in the transition from the Node Active to the Node Sleep state. This can be avoided by evaluating the time a peer link is active. Peer links active less than the Peer Link update Timer plus the time of successive beacon loss should be neglected when evaluating the sleep condition. The minimum time a peer link has to be active was therefore set to 40 sec in the simulation. The maximum packet loss was set to 25 packets. This means that peer links to a deactivated MSTA of an active communication link are closed after 0.5 sec when using a packetization rate of 50 pps according to the VoIP parameterization. A loss of at least 25 packets additionally to the packet loss due to the re-routing time, if a re-routing over another MSTA is possible, is the result.

Table 6.4 shows those simulation parameters for the dynamic network simulations which are different from or in addition to the ones of the static network simulations. The initial random active time was again set to uniform $(5, 25)$ min as used in Section 5.2.3. After a change to the states Node Active or Node Sleep the timers SAT or SST are reset by a random time out of the interval uniform $(8, 24)$ min.

Table 6.4: Simulation parameters that are in addition or have different values in the ns-3 dynamic scenario than in the ns-3 static scenario (see Tab. 6.2).

Sleep-wake algorithm	
Initial active Time	uniform [5,25] min
Station Active Timer	uniform [8,24] min
Station Sleep Timer	uniform [8,24] min
Peer Link update Timer	15 sec
Communication Link	
Start Time1	uniform [20.5,21] min
Stop Time1	uniform [23,23.5] min
Start Time2	uniform [78.5,79] min
Stop Time2	uniform [81,81.5] min

6.2.2 Dynamic Network Simulation Results

Table 6.5 shows the results of the dynamic simulation for T_1=22 min and T_2=80 min. Simulations at later time instances were not carried out. The main reason is the long simulation times. However, it is also assumed that the results which could be obtained at later time instances would be close to those obtained by the static simulations due to a low congestion when most of the MSTAs are already out of power.

The simulations for T_1=22 min and T_2=80 min will show the influences of the protocols on the settling process of the algorithm. The results show a reduction of successful calls of about 3% to 12% compared to the static simulations. However, for the configuration with NoN$_{Th}$ = 7 or 8 the probability of a successful call is still higher than 94%. The bar graph in Fig. 6.2 shows the comparison of the static and dynamic ns-3 simulation results for T_2=80 min. The mean number of active MSTAs which is shown in Tab. 6.3 is calculated in an interval of $\pm$90 sec for all simulation runs. Thereby MSTAs which are shut down during the calltime, i.e. T_1 or T_2 $\pm$90 sec, are treated as not active. When looking at the mean number of active MSTAs we find results of about two stations less than in the static scenario, which is explained as follows: activation and deactivation of non-user MSTAs is based on all of the 102 MSTAs. The two user

MSTAs thus are part of the simulation as well and might be neighboring MSTAs of several relay MSTAs themselves. This is in contrast to the Matlab simulation in Section 5.2.3 where the two user MSTAs are not part of the simulation. This explains the result of the mean number of active MSTAs which is about two MSTAs less when compared to the static network simulation results shown in Tab. 6.3.

Table 6.5: Probability of successful calls – dynamic setup.

	Ratio of Successful Calls	mean number of active MSTAs
NoN_{Th}=6, T_1=22 min	86.1%	45.46
NoN_{Th}=7, T_1=22 min	94.2%	51.19
NoN_{Th}=8, T_1=22 min	96.3%	56.60
NoN_{Th}=6, T_2=80 min	87.8%	52.12
NoN_{Th}=7, T_2=80 min	95.4%	57.00
NoN_{Th}=8, T_2=80 min	98.0%	61.76

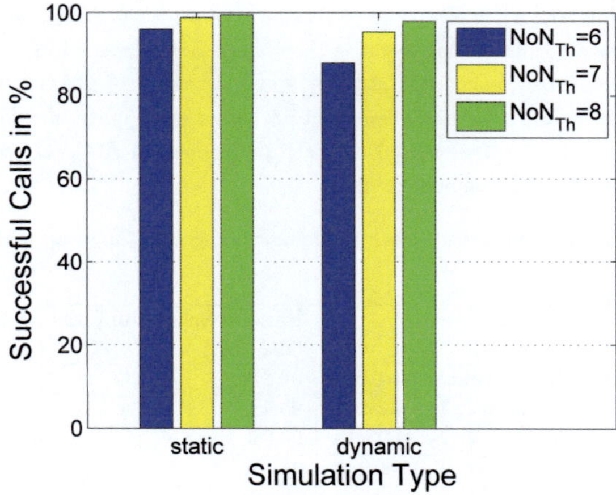

Figure 6.2: Comparison of the ns-3 results for the static and dynamic network simulation for the time instance T_2=80 min.

6.3 Comparison of Different QoS Access Categories

So far all simulations have been carried out with the default QoS access category best effort (AC_BE) in ns-3. However, for a VoIP stream the access category for Voice (AC_VO) might be used by default. The differences of the two access categories are shown in Tab. A.1. This results in the following changes of parameters: AIFSN=2, CWmin=3 and CWmax=7. Because of the lower CWmin and AIFSN value the packets are prioritized compared to the other access categories when contending for the channel access.

Table 6.6 shows the simulation results as obtained by the static network simulations when using the QoS access category for voice (AC_VO) compared to the previously obtained results shown in Tab. 6.3 for the QoS access category best effort (AC_BE) and sleep-wake parameter $\text{NoN}_{Th} = 7$ at T_1=22 min, T_2=80 min, T_3=300 min and T_4=600 min. The results obtained with AC_VO show a degradation of about 15% when compared to the initial configuration at T_0=0 min and about 10% when compared to the static network simulation setup at T_1=22 min, T_2=80 min and T_3=300 min. The reason is the change of CWmin which leads to only four backoff slots in the first transmission attempt and the change of CWmax which leads to a maximum of eight backoff slots and therefore to a higher probability of collisions. Therefore as an important result it can be stated that the parameters of AC_VO are not useful in a multihop scenario for VoIP communication.

Table 6.6: Probability of successful calls – static network simulation setup with different access categories.

	Ratio of Successful Calls	
	(AC_BE)	(AC_VO)
NoN_{Th}=n.a., T_0=0 min	98.8%	83.4%
NoN_{Th}=7, T_1= 22 min	99.4%	88.7%
NoN_{Th}=7, T_2= 80 min	98.8%	87.4%
NoN_{Th}=7, T_3=300 min	96.8%	86.2%
NoN_{Th}=7, T_4=600 min	56.5%	51.2%

6.4 Conclusion

The performance of the DRS mesh network when using the distributed sleep-wake algorithm has been validated by ns-3 simulations of VoIP calls between individual rescuers. First, the network performance has been evaluated in a static setup at dedicated instants in time and thinned out network according to the proposed sleep-wake algorithm carried out in Matlab system simulations. The probability of a successful call thereby is about 99% for all configurations with PREQ messages sent as broadcast. Second, the network has been evaluated in a dynamic setup with full execution of the proposed sleep-wake algorithm at each MSTA individually during the ns-3 simulation. By fully considering the additional network traffic which is a result of the dynamic execution of the algorithm the successful call probability is slightly reduced by 3% to 12% depending on the configuration. When using the configuration with threshold of $NoN_{Th} = 7$ neighbor nodes the probability of a successful call is still about 95%.

Chapter 7

Hardware Requirements on the Air Interface

Contents

In network simulations a correct representation of the physical layer is essential in order to achieve reliable results which are comparable to real hardware performance. On the other hand, this representation of the physical layer needs to provide a relatively high level of abstraction in order to keep simulation times short. The network simulator ns-3 specifies two error rate models, introduced in Section 3.2.3, for the calculation of the bit error rates and corresponding packet error rates for orthogonal frequency division multiplexing, the YANS [88] and NIST [89] error rate models. Section 3.2.3 showed a validation of both models and proposed changes which have been compared with an exact physical layer simulation using Matlab. The recommended changes will be turned into a new model described in Section 7.1.1 and will be further compared with measurements over an AWGN channel with a typical IEEE 802.11 a/b/g/n wireless LAN module. Thereby the new model leads to results that are consistent with exact simulations of the IEEE 802.11a PHY and being comparable to real hardware performance.

Another critical situation in wireless mesh networks occurs when a node in the chain of an ongoing link fails, e. g. because of poor wireless signal conditions or if a node leaves the mesh network or is shut down which is also the case when using LEA. When the link failure is recognized a new path has to be determined by the path selection protocol HWMP. The time which is introduced from the link failure to the first transmitted packet over the new path is commonly called re-routing time. This time will be analyzed by testbed measurements and further compared to network simulations and the influence on VoIP will be shown.

Finally, the influence of processing delays on the VoIP performance will be analyzed. Processing delays will be estimated by an appropriate testbed setup and included in the network simulator by an appropriate model. To this end a Voice over IP (VoIP) communication link is evaluated in a linear chain topology of several nodes within a hardware testbed and comparison to simulation results is made.

7.1 Estimation of the Physical Layer Performance

7.1.1 New ns-3 Error Rate Model

So far the ns-3 YANS model which is using the union bound (Eqns. (3-38) and (3-39)) and the ns-3 NIST model which is using the Chernoff bound (Eqns. (3-48) and (3-49)), detailed in Sections 3.2.3.2 and 3.2.3.3, are using an inaccurate calculation of the corresponding SNR and a wrong calculation of the PSR by an inadequate number of bits L (Eqns. (3-41) and (3-50)). The new error rate model developed in this thesis is using the upper bound (Eq. (3-56)), a correct representation of the SNR (Eq. (3-55)) and the correct number of data bits for the calculation of the PSR according to Eq. (3-50). The proposed changes have been implemented in the network simulator in a new error rate model and have been published in [21] and [22] by the author of this thesis.

Figure 7.1 shows the results of the calculation of the upper bound and Chernoff bound for $K = 1$ and $K = 10$ in Matlab and the exact Matlab simulation of the PSR with BPSK for $R = 1/2$ over an AWGN channel when compared to the new ns-3 model for BPSK ($R = 1/2$). As expected, the Matlab implementation of the upper bound with $K = 10$ (red solid line) corresponds to the ns-3 implementation of the new error rate model (black dash-dot line) and validates the implementation in the network simulator. The proposed changes are also valid for other modulations and code rates which are not presented here.

Finally, the implementation of the new error rate model in ns-3 proposed here which is based on the calculation of the upper bound with $K = 10$, showed a computational overhead of about 14% compared to the NIST implementation. The Chernoff bound even at $K = 1$ shows a deviation of less than $1\,\mathrm{dB}$ when compared to the upper bound and can be used in time consuming simulations when using the proposed changes shown in Section 3.2.3.4 with the hint that the results are slightly optimistic compared to the upper bound. The ns-3 simulations have been carried out by sending 81.000 packets.

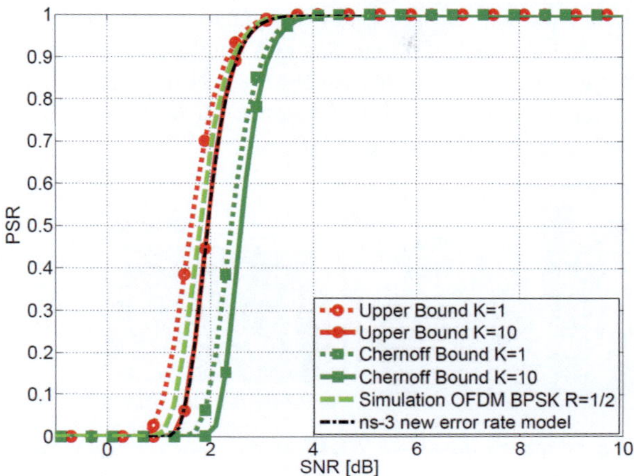

Figure 7.1: Result of the ns-3 implementation of the new model when compared to the upper bounds as well as to the exact Matlab simulation of the PSR with BPSK for $R = 1/2$ over an AWGN channel.

7.1.2 WLAN Hardware Module

The measurements in this work are based on WLAN modules which have been developed during this work together with the SME *embedded wireless GmbH* as part of the FP7 research project CHAMELEON (Project reference: 286510). The focus of this research project was the development of a cost-effective, fast-to-deploy, low-power and flexible video surveillance system that automatically combines images from multiple cameras to create a 180° panoramic view where video data is transmitted via a low-power, small size IEEE 802.11a/g/n Wireless LAN card enabling flexible wireless installation scenarios. Each WLAN module consists of a Qualcomm Atheros QCA9344 System-on-Chip (SoC) which allows dual-band (2.4/5 GHz) operation with up to two spatial streams. Figure 7.2 shows the peripheral board and the WLAN module in detail which is connected by a PCI express interface. The peripheral board provides three Ethernet ports, power supply and two USB ports. The USB-B port is used to connect the module to a PC or power supply. By using an FTDI chip and Virtual COM Port (VCP) driver the USB device appears as an additional COM port available to the PC and can be accessed by application software. The USB-A port can be used to connect external storage devices. Figure 7.3 shows the block diagram of the QCA9344 SoC. The peripheral board is using only three of the five Ethernet interfaces.

The QCA9344 SoC consists of a 74Kc MIPS processor with a high single thread performance operating at up to 533 MHz. Further the WLAN module consist of 16 MB Flash Memory where the firmware is stored. The open-source GNU/Linux distribution for embedded devices OpenWrt [16] is installed on each of the modules with the Version Chaos Calmer (SVN release 45514). OpenWrt allows the setup of different IEEE 802.11 modes, e.g. Client, Access Point or Mesh STA. In this work each module is configured as Mesh Station with the same mesh ID and channel based on the WLAN-stack mac80211.

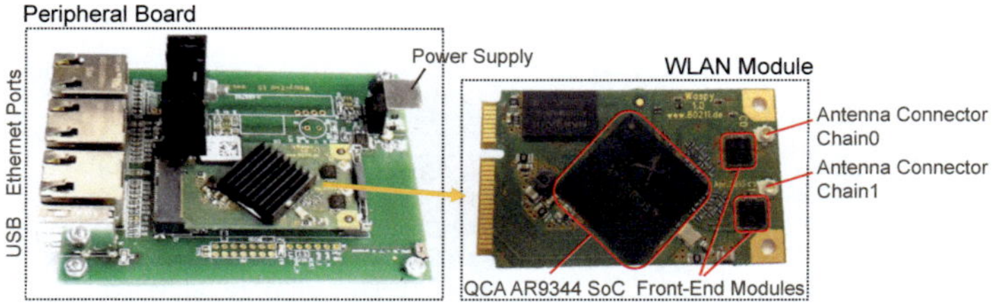

Figure 7.2: Peripheral Board and WLAN Module based on QCA AR9344 SoC.

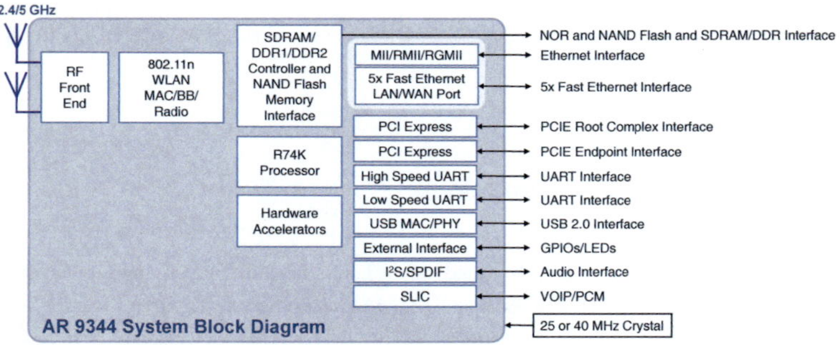

Figure 7.3: QCA AR9344 block diagram [119].

The QCA9344 WLAN MAC consists of 10 Queue Control Units (QCUs) and 10 Distributed Coordination Function (DCF) Control Units (DCUs), a single DMA Rx unit (DRU) and a single Protocol Control Unit (PCU) [119]. Each QCU is linked to exactly one DCU. Ready frames are passed from a QCU to its associated DCU. The DCU manages the Enhanced Distributed Coordination Function (EDCF) channel access procedure on behalf of the QCUs associated with it. The 10 DCUs implement the EDCF channel access arbitration mechanism defined in IEEE 802.11e. This is in contrast to the standard IEEE 802.11 [35] as it is stated that the EDCA mechanism defines only four access categories (ACs) that provide support for the delivery of traffic with user priorities levels (UPs) at the QSTAs[1] [35]. The four transmit queues and the four independent Enhanced Distributed Channel Access Functions (EDCAFs), one for

---

[1]The term QSTA is used in wireless networks for stations supporting the IEEE 802.11e Quality of Service standard.

each queue, are also described in Section 2.2.2 and Appendix A. When supporting the EDCA mode of operation, the EDCAF is responsible for the control of the data streams for each of the four defined access categories. Within the QCA AR9344 SoC eight DCUs map directly to the eight EDCF UPs. The two additionally DCUs handle beacons and beacon-gated frames. The mapping of physical DCUs to absolute channel access priorities is fixed and cannot be altered by software (see [119] for further information). The network simulator ns-3 provides only four transmit queues and four independent EDCAFs as intended by the standard [35]. For the use case considered in this work with a low number of data streams of a single access category this difference does not matter.

7.1.3 Estimation of Hardware Parameters

The packet success rate (PSR) of the WLAN module over an AWGN channel can be measured with one module configured as transmitter and one module configured as receiver by using a wired connection with a variable attenuator (see Fig. 7.4).

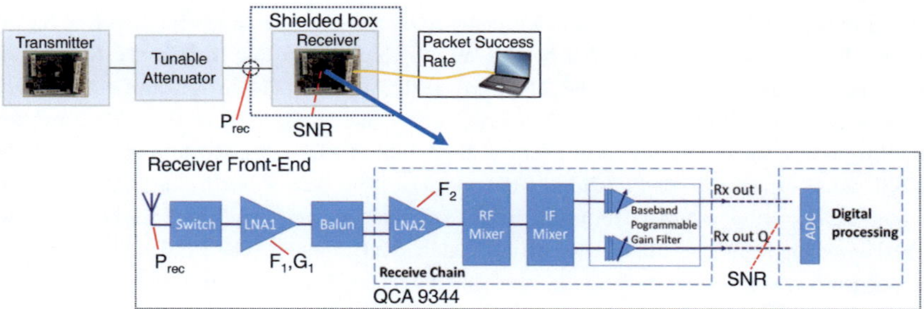

Figure 7.4: Measurement setup and WLAN receiver block diagram.

The Atheros radio test software provided by chip manufacturer Atheros for supporting Wi-Fi qualification measurements can be used for radio evaluation and manufacturing tests. Various transmission tests, receive and link test and calibration through manufacturing flow can be performed. The link test which is used here requires that two stations are running the software. One station transmits a dedicated number of packets at a predefined data rate (with possible values from 6 Mbps to 54 Mbps for purposes of this thesis) and the other station receives and displays statistics including the packet error rate and received signal strength indication (RSSI). The WLAN receiver module is placed in a shielded box. Figure 7.4 shows also the receiver block diagram. The module consists of an external SE5516A Front-End-Module (FEM) from SKYWORKS [120]. The SE5516A is a dual-band 802.11 a/b/g/n/ac WLAN FEM which consists of a power amplifier for 2.4 and 5 GHz, transmit filter, diplexer, transmit/receive switches and low noise amplifier for both frequency bands. The noise figure NF_1 and gain G_1 of the external low noise amplifier including the switch of the FEM for the 5 GHz band, named as LNA_1, are shown in Table 7.1. Measurements of an evaluation board of the FEM confirmed the values shown in Tab. 7.1. The noise figure NF_2 is associated with the internal low noise amplifier LNA_2 of the QCA AR9344 SoC. The LNAs and PAs of the FEM are connected by discrete baluns (balanced to unbalanced converters) to the differential ports of the chip.

Table 7.1: Noise Figure and Gain of WLAN module.

Hardware Parameters	
NF LNA_1	$NF_1 = 2.8\,\mathrm{dB}$
Gain LNA_1	$G_1 = 12.0\,\mathrm{dB}$
NF LNA_2	$NF_2 = 5.8\,\mathrm{dB}$
Overall Noise Figure of Rx Chain	$NF = 3.185\,\mathrm{dB}$

In order to compare the measured PSR to the simulation results, the SNR at the A/D converter in the WLAN receiver has to be estimated. One possibility is to use the Received Signal Strength Indicator (RSSI) provided by the WLAN chip. According to [35], the RSSI is an optional parameter that has a value of 0 through RSSI Max. This parameter is a measure provided by the PHY sublayer and is indicating the energy observed at the A/D converter used to receive the current Physical Protocol Data Unit (PPDU). An absolute accuracy of the RSSI values reported by the modules is not specified [35]. [121] shows a practical conversion from RSSI to dBm values. The mean RSSI from different vendors have been compared in [122] revealing significant differences between individual vendors. The RSSI reported is also dependent on design choices made by the manufacturer. For more advanced chipsets using IEEE 802.11n multiple-input/multiple-output techniques this becomes even more challenging and the calculation of the RSSI is not published by the vendors. Because of these inaccuracies the measurement of the RSSI is not practicable to evaluate the PSR. Another approach is the measurement of the signal power at the receiver input and calculation of the SNR at the ADC by subtracting the calculated noise figure of the LNAs. For this purpose, an exact knowledge of the RF front-end design is required.

The SNR at the A/D converter in the WLAN chip (Fig. 7.4) can be calculated by subtracting the noise power N_p and the noise figure NF of all components in the signal chain from the measured received power $\mathcal{P}_{\mathrm{rec}}$ at the antenna connector.

$$\mathrm{SNR}\,[\mathrm{dB}] = \mathcal{P}_{\mathrm{rec}}\,[\mathrm{dBm}] - N_p\,[\mathrm{dBm}] - \mathrm{NF}\,[\mathrm{dB}] \tag{7-1}$$

With the noise power N_p

$$N_p = 10\log\left(\frac{k_B T B}{1\mathrm{mW}}\right) \tag{7-2}$$

and the over all Noise Figure NF

$$\mathrm{NF} = 10\log(F) \tag{7-3}$$

$$F = F_1 + \frac{F_2 - 1}{G_1} \tag{7-4}$$

k_B is the Boltzman constant, T the absolute temperature and B the noise Bandwidth. F_1 and G_1 are the noise factor and gain of the external FEM (LNA$_1$ Tab. 7.1). F_2 is the noise factor of the internal LNA (LNA$_2$ Tab. 7.1). The packet size in the measurements and for the calculation of the PSRs is 1000 bytes.

7.1.4 AWGN Channel Measurements

Figure 7.5 shows the result of the measurement over an AWGN channel (blue dash-dotted line) compared to the upper bound and Chernoff bound for $K = 1$ and $K = 10$ as well as the ns-3 implementation of the new model proposed in Section 7.1.1.

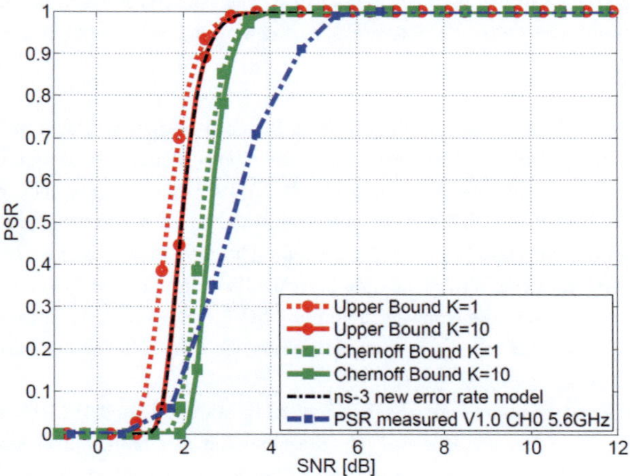

Figure 7.5: Result of the ns-3 implementation of the new model when compared to the Upper Bounds and PSR of BPSK for $R = 1/2$ over an AWGN channel.

The measurements are carried out by using IEEE 802.11a at a frequency of 5.6 GHz for BPSK with code rate 1/2. Only Chain0 of the two available antenna ports of the WLAN module is shown. The results show a good correspondence of the measurement results with the upper and Chernoff bound for $K = 10$ as well as with the ns-3 implementation of the new model especially for low PSRs. With the changes described in the previous sections land including the noise figure of the WLAN hardware module in the network simulation, the YANS and NIST model – when using a correct SNR calculation and correct number of bits in ns-3 – could be used as well to estimate the performance of the physical layer of a typical WLAN module. In this case YANS would correspond to the union bound with $K = 1$ and NIST would correspond to the Chernoff bound with $K = 10$. With the hint that as seen in Fig. 3.16 the corrected version of the YANS model would still show too optimistic results of about 1 dB and the NIST model would show pessimistic results by less than 1 dB compared to the new model. Both, the corrected NIST and the corrected YANS model could be used.

The flattening of the measured AWGN graph (blue dash-dotted line) at high packet success rates might be due to inaccuracies in the measurement setup, where only the WLAN receiver module is placed in a shielded box. Part of the signal from the transmitting module might be propagating over the air and feeding back on the screen of the cable, finally entering at low amplitude the screened box where the receiver is placed. This results in a slightly frequency selective fading characteristic. To prevent this the transmitting module needs to be placed in a shielded box as well and cable lengths should be as small as possible. A more plausible explanation is that the fading characteristic seen at the measurements is a result of a mismatch of the ports which leads to reflections on the connected signal lines. Therefore essential for the validation is the region of low packet success rates. The sensitivity which is defined at a PSR of 0.9 corresponds to the received Power P_{rec} at the antenna connector and is about -93 dBm for BPSK (6 Mbps).

Figure 7.6 shows additional measurement results for both frequency bands at 2.412 GHz and 5.6 GHz and for both Chains. The measurements have been carried out using a LitePoint radio tester. This Equipment consists of a golden unit WLAN signal generator. The WLAN receiver module has been placed in a shielded box and connected to the radio tester and the measurements are based on 500 packets per configuration. The measurements have been carried out with a newer version of the WLAN module (version 1.1). In addition to some minor changes of some GPIO connections the front-end module (FEM) has been changed as well, where the characteristics of the FEM are still the same as shown in Tab. 7.1. Further information about the characteristics can be found in [123].

The measurement at Channel 1 for the IEEE 802.11g standard at 2.412 GHz for Chain0 fits exactly to the ns-3 simulation. However, the measurement at Chain1 shows also slight fading characteristics. The measurement at 5.6 GHz at Chain0 shows comparable results to the previously obtained results with the WLAN module with version 1.0 even though the front-end module has been changed. The results of Chain1 show also fading characteristics and slightly better results. The fading characteristic seen at the measurements, as stated earlier, might be a result of a mismatch of the ports which leads to reflections on the connected signal lines with the best match shown at 2.412 GHz for Chain0. The deviation of the measurement results for the different chains and frequencies is about 2 dB which is in the same range of deviations of the original ns-3 YANS and NIST model as shown in Section 3.2.3.4. The YANS model showed overly optimistic results of about 2 dB whereas the NIST model showed pessimistic results of about 2 dB. The influence on network simulation results and impact of the accuracy on the models will be further analyzed in the following section.

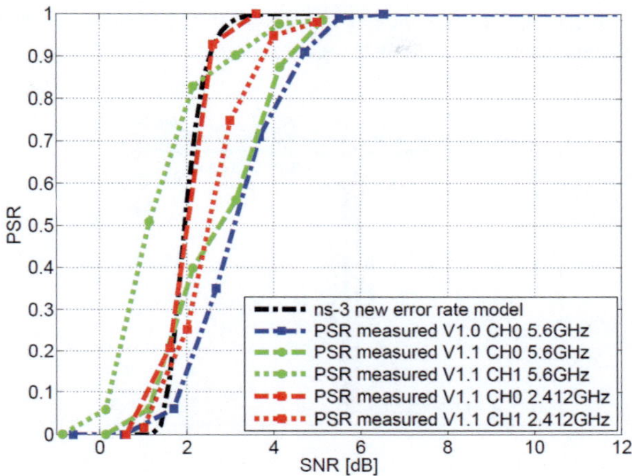

Figure 7.6: Result of the ns-3 implementation of the new model when compared to the PSR of BPSK for $R = 1/2$ over an AWGN channel for 2.412 GHz and 5.6 GHz.

7.1.5 Evaluation of the Influences on Network Simulation Results in ns-3

When looking on very specific network configurations even variation of a few dB might have strong impact on network performance. This will be shown by the following example, where it should be noted that for most network configurations the impact is much weaker.

Showing this importance of a correct abstraction of the physical layer, the influences on network simulation results of the three models (YANS, NIST and the New Model) are evaluated in an ns-3 simulation covering a hidden node scenario as an example of a typical WLAN network problem. The evaluation is carried out with four communication links, involving five nodes which are placed on a rectangular grid. Figure 7.7 shows the topology in ns-3 where distances are in meter. Each node is initialized as a Mesh Station (MSTA) based on the IEEE 802.11s standard [35] and is part of a mesh network. Each MSTA which is in range of another MSTA establishes a peer link in between them and allows the routing of data throughout the stations.

In Fig. 7.7 the communication links from Node1 to Node3 (connection 1) and Node3 to Node1 (connection 2) are relayed by Node2 because no direct link is possible. A relaying over Node4 is also possible which is decided by the routing process. Node4 and Node5 are connected by a direct link (connection 3 and connection 4). Also some of the nodes are hidden to others. Each communication link has a duration of 100 sec and each connection is modeled as UDP data stream with 500 kbps. Connection 1 starts at a simulation time of 2 sec, connection 2 at 2.1 sec, connection 3 at 10.2 sec and connection 4 at 10.3 sec. With the packet payload of 922 bytes, IP and UDP headers the packet becomes 1000 bytes in total.

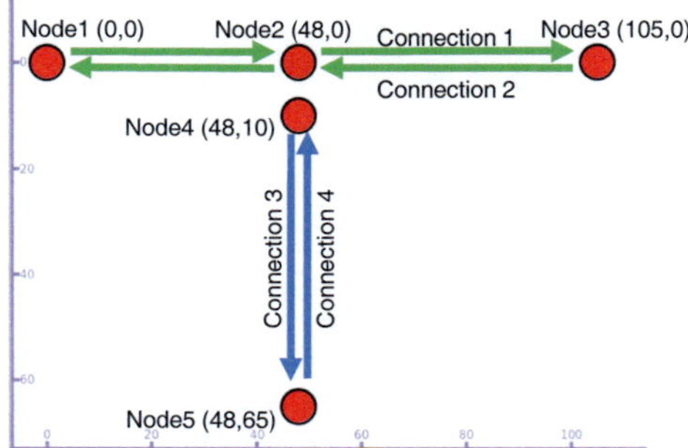

Figure 7.7: Topology of the simulation in ns-3 with 5 nodes and 4 connections.

For the calculation of the received signal power at the individual nodes, the ns-3 Log-Distance-Propagation-Loss-Model [83] is used with an exponent of 2.7 and the Reference Loss calculated at a center frequency of 5.6 GHz. The transmit power is chosen to be 0 dBm and the noise figure is 3.185 dB. Table 7.2 shows the resulting SNR values and PSRs for the different distances d_t of connections 1 to 4 of the three models based on the simulation parameters for different transmission distances of each of the nodes for BPSK coderate $1/2$. $d_t(i)$ denotes the Euclidean distance between two individual nodes.

Table 7.2: Results of the PSR for different distances d_t.

Distance $d_t(i)$	48 m	55 m	57 m	65 m
SNR	4.973 dB	3.377 dB	2.958 dB	1.418 dB
YANS PSR in %	100	100	99.99	97.95
NIST PSR in %	99.72	35.02	1.57	0
New Model PSR in %	100	99.47	97.61	2.77

Figure 7.8 shows the mean packet success of each individual connection of the three models (YANS, NIST and the New Model) with 1000 simulation runs. The different error rate models yield different PSRs for a given SNR, resulting in different transmission ranges as seen in Tab. 7.2 and the specified scenario is highly sensitive to this fact. The YANS model has a packet success of almost 100% for all connections. With the NIST model almost no transmission is possible with connection 1 and 2 and the New Model has a packet success of about 82% for connection 1,2 and 98% for connection 3,4. This shows that the network simulation results might be highly sensitive to the specified error rate model as it is the case in this special topology. Considering the measurement results of the previous section with similar deviation

may lead to similar results in the network simulation. However, when using a not correct physical layer model the results will be even worse. This shows also that an abstraction of the physical layer avoiding any unnecessary inaccuracies is key for reliable simulation results on network level. Of course the topology used in the simulations with nodes placed in the range of the maximum transmission range is highly sensitive to minor variations of the SNR. Therefore simulation topologies must be chosen carefully and models have to be parametrized properly.

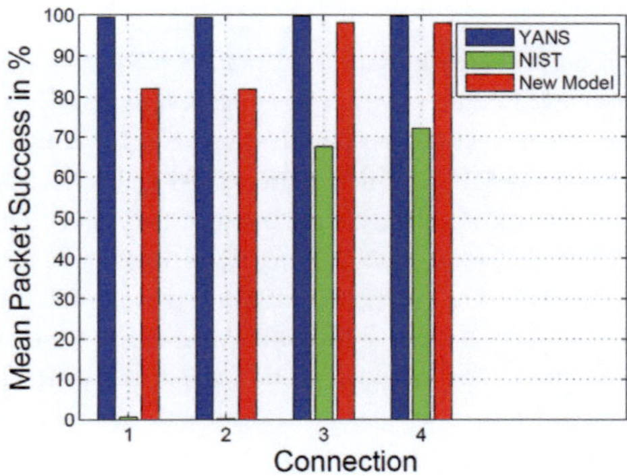

Figure 7.8: Mean packet success in % depending on the configuration.

7.1.6 Conclusion

The simulation results of the new ns-3 error rate model in Sec. 7.1.1 show a good correspondence with the measurement results of a typical WLAN module (Sec. 7.1.4) as well as with an exact physical layer simulation in Matlab of an IEEE 802.11a OFDM communication link (Sec. 3.2.3.4). Key elements of the new model are the usage of the upper bound (Eq. (3-56)) with $K \geq 10$, correctly taking into account the SNR at the receiver input (Eq. (3-55)) and correctly taking into account the number of data bits used for the calculation of the PSR in Eq. (3-50). Because of the inaccurate SNR and PSR calculations in YANS and NIST the YANS model is overly optimistic (about 2 dB) whereas the NIST model is too pessimistic (about 2 dB) when compared to the new error rate model. The calculation of the SNR should be changed in both, the ns-3 NIST and YANS error rate models in order to correctly take into account the error rate in the OFDM system. The number of data bits used in the calculation of the packet success rate should be changed as well in order to take the convolutional coding into account correctly. With the proposed changes both models can be used, resulting in more reliable performance estimations of the PSR. In order to get reliable results in the ns-3 simulations, it is crucial to take into account the noise figure of the RF-Front-End. Finally it has been shown, that the network simulation results may be highly sensitive to the specified error rate model which at least for some network scenarios shows that an abstraction of the physical layer avoiding any unnecessary inaccuracies is key for reliable simulation results on network level.

7.2 Re-Routing Performance Evaluation

In heterogeneous mesh networks with static and mobile MSTAs it may happen that some MSTAs leave or join the network during data transmission. The described sleep-wake algorithm in Chapter 5 also allows the activation and deactivation of MSTAs. If a MSTA is running out of power or leaves abruptly the MBSS an active link is lost and a re-routing of the connection is necessary. Thereby it has to be distinguished between two cases. On one hand, a MSTA which is leaving the MBSS will sent Mesh Peering Close messages in order to inform all its neighboring MSTAs. On the other hand, if a MSTA leaves abruptly the network no Mesh Peering Close messages are sent. An active link will experience a large number of lost packets before detecting the lost link. The following section shows the analysis of measurements of a re-routing scenario carried out in [124] and further the comparison and evaluation with a network simulation in ns-3.

7.2.1 Testbed Evaluation of a Re-Routing Scenario

7.2.1.1 Methodology

Figure 7.9 shows the setup of the routing scenario with four MSTAs. The MSTAs A and B are the end nodes of the communication link with no direct connection. The datastream is routed over MSTA C or MSTA D.

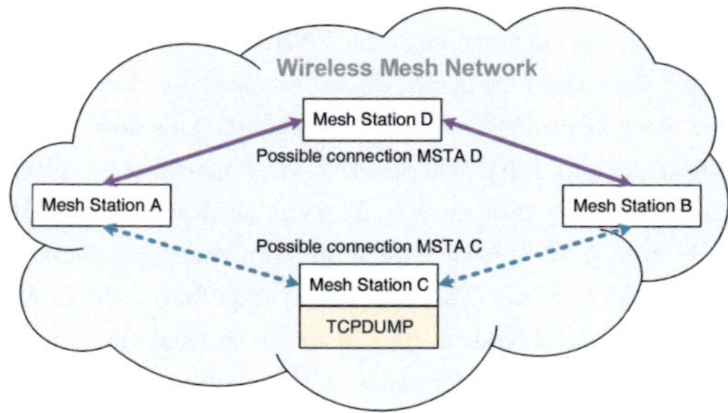

Figure 7.9: Measurement setup of the re-routing scenario [124].

The WLAN modules as described in Section 7.1.2 have been used and placed in the university lab according to Fig. 7.9. The maximum possible datarate was adjusted to 6 Mbits/s (BPSK with coderate 1/2) on each of the modules. In this way no adaptive adjustment of the datarate is used and the datarate remains the same during the measurements. The communication between the two end nodes is realized with the software tool Iperf [125]. Iperf can be used for active measurements of the maximum achievable bandwidth on IP networks. It supports various protocols (e. g. TCP or UDP with IPv4 and IPv6). The bandwidth, loss, delay jitter and other parameters are reported on each test. Iperf can be installed in OpenWrt and executed on command line. A constant bit rate UDP stream is created from MSTA A to MSTA B and vice versa which emulates a voice communication over the network. Both of the intermediate nodes (MSTA C and MSTA D) are mutually switched on and off causing a routing or re-routing between the end nodes. The Wi-Fi interfaces will be turned on and off 25 times on each measurement run through appropriate shell scripts.

In addition, TCPDUMP [126] is used on one of the two middle nodes in order to log the air interface and store protocol captures in a pcap trace file. TCPDUMP is a common packet analyzer which also uses the libpcap library to capture packets. TCPDUMP version 4.5.1 has been installed in the used OpenWrt version. The measurement process is carried out by shell scripts which need to be synchronized on each of the MSTAs. This can be done with the time-based job scheduler Cron. Additionally, time synchronization among the 4 MSTAs is provided, using the Network Time Protocol (NTP). The protocol is already implemented on the WLAN modules and can be used with only a few changes. With NTP time can be kept over the public Internet with an accuracy of 10 ms. In local networks, accuracies in the range of $200\,\mu s$ are possible under ideal conditions [124]. Absolute accuracy is not required in this scenario because TCPDUMP is only used on one of the two middle nodes in order to log the air interface by its configuration as monitoring module. For the synchronisation of the shell scripts the accuracy in milliseconds is sufficient because the waiting times in the shell scripts are in the range of seconds. The Iperf target bandwidth is set to 68.8 kbits/sec. One MSTA is set as client and one MSTA as server with the respective commands shown in Tab. 7.3. The parameter dualtest allows a bidirectional link emulating a VoIP stream with length (len) of 172 bytes payload.

Table 7.3: Initialization of Iperf as server and client for an emulated VoIP-connection with speech-codec G.711.

Server:	`iperf --udp --server --len 172 -fk --interval 1`
Client:	`iperf --udp -c 192.168.1.2 --len 172 --bandwidth 68800`
	`--dualtest -fk --interval 1 --time 540`

Figure 7.10 shows the measurement work flow on the intermediate MSTAs. On MSTA C a Wi-Fi interface is generated which is configured as additional monitor interface. This monitor interface allows the logging of the air interface by TCPDUMP. Then the datarate is set to a fixed value on both of the modules and TCPDUMP is started on MSTA C using the monitor

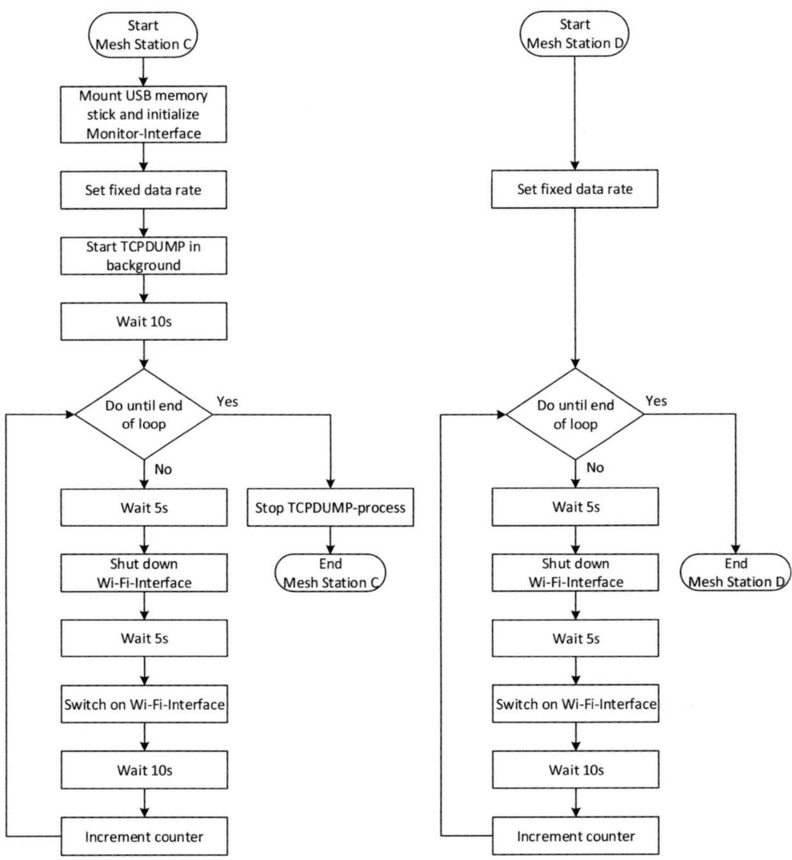

Figure 7.10: Flow chart: Re-routing setup of intermediate MSTAs [124].

interface. In order to alter the on-off phases of both modules, MSTA C is paused for 10 sec. Then the loop starts within which both MSTAs are paused first for 5 sec. After this pause of 5 sec the Wi-Fi interface is shut down and turned off for a period of 5 sec. Before each shut down the MSTA is sending a Mesh Peering Close message to its neighboring MSTAs. After another 5 sec the Wi-Fi interface on the MSTA is switched on again. In that way the modules are active for 15 sec each. The off phase is 5 sec at each MSTA and the overlap period when both modules are active has a length of 5 sec as well. The idea thereby is that the newly activated MSTA has enough time to establish peer links to all its neighbors. When the previously active MSTA is shut down, the routing process to the new active MSTA can immediately start after receiving a Mesh Peering Close frame. After the end of the loop the TCPDUMP process is stopped and the data can be analyzed. Figure 7.11 shows the on-off phases of the intermediate MSTAs C and D. It can also be seen that at the beginning both MSTAs are active and the connection is either routed by MSTA C or by MSTA D. This must be considered when evaluating the results.

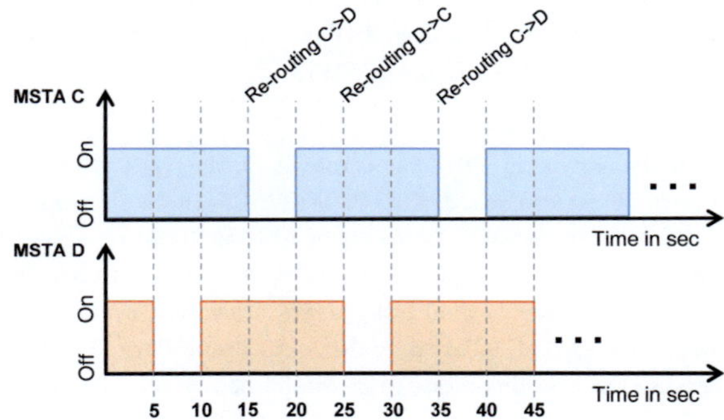

Figure 7.11: On-off phases of the re-routing setup of intermediate MSTAs according to the flow chart in Fig. 7.10.

7.2.1.2 Measurement results of the re-routing scenario

In [124] the measurement results have been analyzed assuming a constant bandwidth of the datastream over the measured time. However, sporadic dips have been observed which might be either a result of varying channel conditions or due to the packet loss during the re-routing process. A deeper analysis is provided in the following by examination of the captured protocol data.

It first should be noted that it cannot completely be excluded that not all packets have been captured by the monitoring module due to the wireless propagation environment. Therefore a misinterpretation of the captured data could not completely be excluded.

By analyzing the captured packets it can be distinguished between four variants of the re-routing which are described in the following paragraph. In the description of the different variants it is assumed that a link is established between the two end MSTAs over the intermediate MSTA C and will be re-routed to MSTA D.

- Variant 1: the datastream is active and is routed via the respective intermediate MSTA, i.e. for the case of our analysis MSTA C. A peer link to the alternative MSTA D has already been established. According to Fig. 7.10 the Wi-Fi interface of MSTA C will be shut down. Before the shut down, MSTA C is sending a Mesh Peering Close message to its neighbor MSTAs. The sender MSTA receives the Mesh Peering Close and starts the routing with the alternative MSTA. This is the normal operation intended by the flow chart in Fig. 7.10.

- Variant 2: the datastream is active and is routed via the respective intermediate MSTA, i.e. for the case of our analysis MSTA C. A peer link to the alternative MSTA D has already been established. The sender MSTA starts a new path request before the Wi-Fi interface of MSTA C will be shut down. The link is routed over the alternative MSTA, e.g. because of better link conditions. MSTA C will be shut down and is sending a Mesh Peering Close message to its neighbor MSTAs. However, the link has already been re-routed to the alternative MSTA D before MSTA C was shut down. This leads to a quite low number of lost packets.

- Variant 3: The datastream is active and is routed via the respective intermediate MSTA, i.e. for the case of our analysis MSTA C. MSTA C will be shut down and is sending a Mesh Peering Close message to its neighbor MSTAs, however, a peer link to the new MSTA could not yet be established. The peer link establishment procedure was cancelled e.g. due to interference and several lost packets. Now first a new attempt of the peer link establishment is started, followed by the routing procedure. The routing time is now extended because the peer link needs to be established before the routing can start.

- Variant 4: From the results it could also be analyzed that the re-routing took a much longer time because of a non-received Mesh Peering Close frame. The data packets were still sent to the shut down MSTA, however, they were not received and also not acknowledged. A path error then indicates the lost path after several lost packets and a new routing procedure is started.

A further analysis of the captured packets is realized using a post-processing routine with TShark [82]. TShark is a terminal oriented version of Wireshark designed for capturing and displaying packets and supports the same options as Wireshark. Using several display filters for specific management messages, e.g. Mesh Peering Close, Path Reply (PREP) or data packets with dedicated receive and transmit addresses, these filtered data sets can be used for estimating the re-routing time in a post-processing in Matlab.

In a first attempt the re-routing time is calculated by the difference of the first Mesh Peering Close message and the last PREP message. However, in some cases the routing is done before the Mesh Peering Close (variant 2) resulting in a negative routing time during the evaluation. Another approach, which is used here, is to define the re-routing time as the time difference between the last data packet sent to the intermediate MSTA before the link is closed and the first data packet sent to the new intermediate MSTA after the new path has been established. In that way the complete interruption between consecutive data packets is considered.

Figure 7.12 shows the histogram of the re-routing time for the measurement setup shown in Fig. 7.10 with 25 on-off procedures carried out three times. The overall re-routing time is defined, as described above, as the time difference between the last data packet sent to the intermediate MSTA before the link is closed and the first data packet sent to the new intermediate MSTA after the new path has been established. Only variants 1 and 3 can be analyzed when using this evaluation procedure because the required trigger for determining the last data packet sent before re-routing is the Mesh Peering Close frame. In that way only 34 re-routing procedures of the datastream could be analyzed. Variant 2 would result in a negative re-routing time because the routing is done before the peer link has been closed. Also variant 4 cannot be analyzed because of the missing Mesh Peering Close frame. The minimum time measured with this approach was 13.5 ms which is below the packet repetition rate of 20 ms.

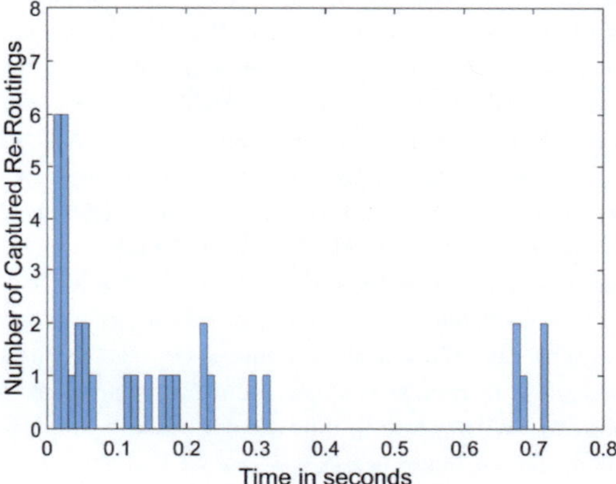

Figure 7.12: Histogram of overall re-routing time of testbed measurements.

7.2.2 Network Simulation of a Re-Routing Scenario

7.2.2.1 Methodology

Figure 7.13 shows the positions of the MSTAs in the ns-3 visualizer. The MSTAs are placed to form the same setup as shown for the testbed measurements (Fig. 7.9). The network simulations have been carried out using the ns-3 Range-Propagation-Loss-Model as described earlier.

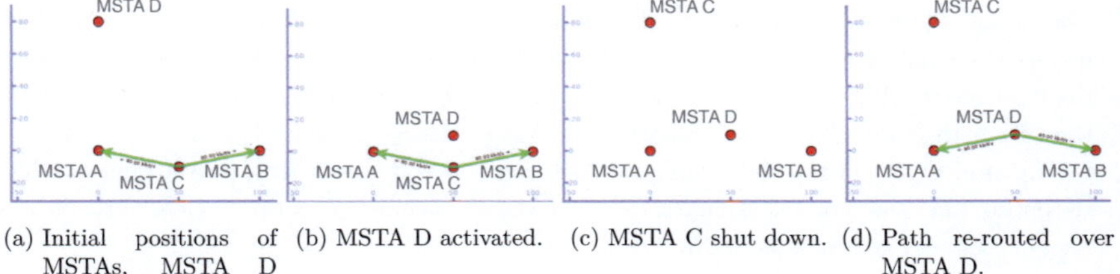

(a) Initial positions of MSTAs, MSTA D deactivated. (b) MSTA D activated. (c) MSTA C shut down. (d) Path re-routed over MSTA D.

Figure 7.13: ns-3 simulation setup for evaluation of the re-routing scenario.

The VoIP call starts randomly for each simulation run within a time interval from 5 to 35 sec after the beginning of the simulation and has a random duration of 120 sec up to 180 sec. The second UDP data stream starts randomly in between 100 ms to 200 ms after the first data stream. The ns-3 visualizer shows only the datastream to the respective last MSTA by a green arrow. Figure 7.13a shows the positions of the MSTAs at the beginning of the simulation. MSTA D is deactivated and out of range of the other MSTAs. The two datastreams are routed over MSTA C. After a certain time which is uniformly distributed in the interval [50,60] sec MSTA D is activated and peer links with the other MSTAs can be established (Fig. 7.13b). After the uniformly distributed time in the interval [80,90] sec MSTA C is abruptly shut down (Fig. 7.13c) and no Mesh Peering Close message is sent[2]. Thus the simulation setup corresponds to variant 4 of the re-routing discussed in the previous section. After the detection of the lost connection the path can be re-routed over MSTA D (Fig. 7.13d).

The ns-3 mesh model implements two ways of the detection of a lost connection. Links can be closed either due to a maximum beacon loss of successive beacons or due to a maximum packet loss of successive packets. If one of the parameters exceeds the threshold the link will be closed. Both parameters will be compared by two individual simulations. First, the maximum beacon loss was set to 20. With a beacon interval of 1.024 sec a link will be closed after at least 20.48 sec. Second, the maximum packet loss was set to 1000 packets. With the packet repetition rate of 50 pps this leads to a duration of 20 sec before the link will be closed as well. Both variants will be compared in the next section.

7.2.2.2 Network simulation results of a re-routing scenario

As already discussed in Section 7.2.1.2 the re-routing time can be defined in two different ways. One definition is that the re-routing time is the difference between the last data packet sent over the previously active path and the first data packet received when a new path has been established as used by the evaluation of the measurement results in the previous section. This time can also be called the overall re-routing time. The overall re-routing time can be easily assessed by evaluating the last data packet in the time interval [80,90] sec when MSTA C is shut down and the first data packet which is sent over MSTA D for both individual data streams. The second definition is, that the re-routing time is only the time difference between the first management message when MSTA C is shut down which should be a Mesh Peering Close to the deactivated MSTA and the last management message which is the last Path Reply before the new path is established to MSTA D. The Mesh Peering Close is sent to the deactivated MSTA after expiry of the threshold condition (max beacon loss or max packet failure). The re-routing time based on management messages thereby is independent of the size of the values max beacon loss and max packet failure.

---

[2]The mechanism of sending a Mesh Peering Close message before shutting down a MSTA is currently not implemented in ns-3.

Figure 7.14 shows the histogram of 100 Monte Carlo simulation runs with a maximum number of successively lost packets of 1000 and Fig. 7.15 the result with a maximum number of successive lost beacons of 20. Both simulations result in an interruption of about 20 sec, the results however show considerable differences. Figure 7.14a shows a dedicated peak at 20.02 ms.

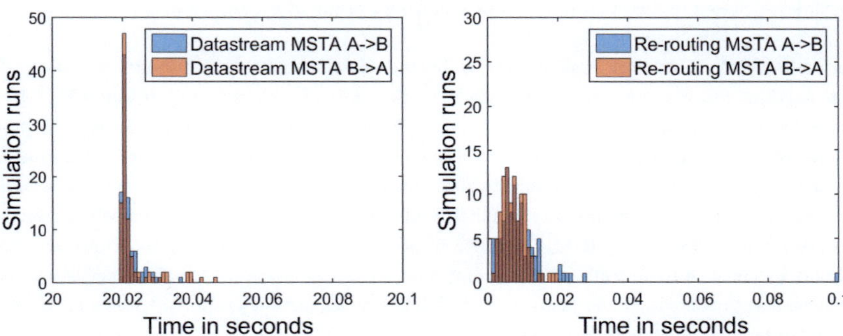

(a) Histogram of re-routing time for data packets. (b) Histogram of re-routing time for management packets.

Figure 7.14: ns-3 simulation result of the re-routing scenario with a maximum number of successive lost packets of 1000.

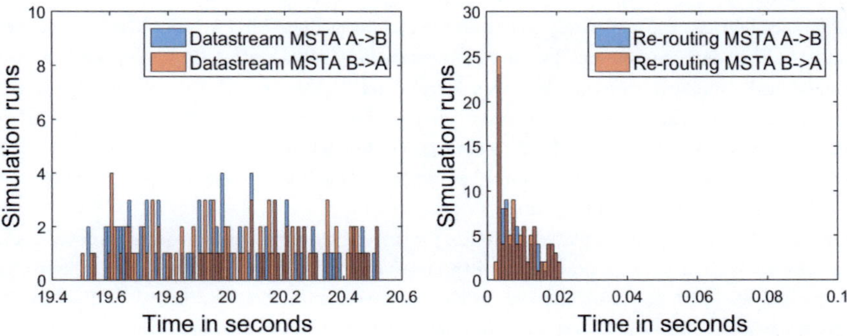

(a) Histogram of re-routing time for data packets. (b) Histogram of re-routing time for management packets.

Figure 7.15: ns-3 simulation result of the re-routing scenario with a maximum number of successive lost beacons of 20.

Figure 7.15a shows a more uniform distribution within a time interval ranging from approximately 19.5 sec to 20.5 sec. If calculated exactly the interval is in between 19 and 20 times the beacon interval of 1.024 sec in addition to the re-routing overhead caused by management traffic.

The evaluation of the re-routing time based on management messages (Fig. 7.14b and Fig. 7.15b) on the other hand leads to comparable results for both setups. The results obtained indicate a low re-routing time for management packets, but a rather long re-routing time for data packets. The long re-routing time for data packets is dependent on the chosen values for the detection of a lost link (max beacon loss or max packet failure). It should be noted that these values have been chosen in a certain range for a simple evaluation of the resulting data.

In the setup of the network simulation with dynamic sleep-wake algorithm shown in Section 6.2 the maximum packet loss was set to 25 packets. Therefore peer links to a deactivated MSTA of an active communication link are closed after 0.5 sec when using a packetization rate of 50 pps. A loss of at least 25 packets in addition to the re-routing time based on management messages is the result. If the beacon loss threshold is used, a re-routing time in the range of several seconds is achieved depending on the number of allowed lost beacons. Of course the beacon interval can be lowered to 100 ms in order to achieve lower re-routing times. However, this would also lead to more traffic and more interference on the network. On the other hand, the beacon loss threshold is more reasonable for MSTAs which are currently not transmitting data and can be used as an additional metric.

7.2.3 Conclusion

The re-routing time estimated by the testbed measurements has been defined as the time difference between the last data packet sent to the intermediate MSTA before the link is closed and the first data packet sent to the new intermediate MSTA after the new path has been established. Similarly this time has been investigated in the network simulation by means of the overall re-routing time. In contrast to the testbed measurements a Mesh Peering Close message is not sent when the MSTA is shut down, but when one of the threshold conditions (max beacon loss or max packet failure) has expired. This leads to long re-routing times when evaluating the overall re-routing time.

On the other hand the simulated re-routing with the threshold condition max beacon loss or max packet failure with evaluation of the re-routing time based on management messages is comparable with the measurement results, however, with more statistical outliers in the measurements. Those can be a result of interference in the wireless propagation environment not modeled in the network simulation. Another explanation might be that the network simulator ns-3 does also not include the effect of additional processing delays due to the software execution time and scheduling mechanism of tasks in the CPU. This effect will be further analyzed in the next section.

The measurement results showed that a low number of re-routings in a VoIP stream over the wireless mesh network are possible while still complying with the requirements for VoIP. This cannot be shown in the simulation because of the missing implementation of sending a Mesh Peering Close message when a MSTA is going to be shut down. The re-routing time is dependent on the threshold condition max beacon loss or max packet failure which leads to re-routing times of about 20 sec in the simulations. In the measurements the Mesh Peering Close is sent before the MSTA is shut down. Of course if a MSTA abruptly leaves the mesh network without sending a Mesh Peering Close message the re-routing time can be much longer and is dependent on a appropriate metric which has been already shown in the simulations. With a maximum packet loss of 25 packets an interrupt of the communication link of 0.5 sec in addition to the management routing time can be achieved. This would lead to an interrupt in the call which might not cover QoS requirements for voice calls, but still be acceptable in a disaster recovery situation if the link is then working again. A mechanism to prevent active MSTAs with an ongoing VoIP stream from shut down could also be foreseen.

7.3 Influence of Processing Delays on the VoIP-Performance

The performance of the proposed lifetime enhancement algorithm in Chapter 5 has been ana-lyzed by a network simulation in ns-3 for VoIP calls between individual rescuers and validated according to the VoIP requirements shown in [105]. Thereby packet loss should be no more than 1 percent, one-way latency (mouth to ear) should be no more than 150 ms and average one-way jitter should be targeted at less than 30 ms. The simulation results showed that for a VoIP communication link of two Wireless Mesh Stations (MSTAs) for a dedicated number of observed scenarios with different node densities more than 90% of successful calls could be achieved. Dropped calls, however, were caused almost only due to packet loss while packet delay and jitter were very low which might not cover the real hardware performance because the Quality of Service (QoS) is dependent on both loss and delay. As most network simula-tors ns-3 focuses on accurate modeling of the MAC- and higher layer protocols. However, the actual simulation model in ns-3 lacks the modeling of appropriate processing delays which are added in each Mesh Station due to the software execution time and scheduling mechanism of tasks in the CPU. So far the introduced delay in the network simulation is caused only by the MAC-layer protocol and duration of the transmission itself as illustrated in the ns-3 node object in Fig. 7.16. The lack of modeling processing delays does also cause some other issues in the network simulation which have been discussed in Section 3.2.2.1.

In Section 3.3 state-of-the-art methodology and possibility to account for the overhead of pro-tocol processing has been discussed. These models are not part of the current ns-3 release as the state of this work. In this section we will focus on a simplified approach where we target to estimate the maximum number of hops possible for a single VoIP connection by including an appropriate processing delay at each MSTA in the network simulation. The processing delays will be estimated for a single state of utilization, i. e. only one setting of traffic load, modulation

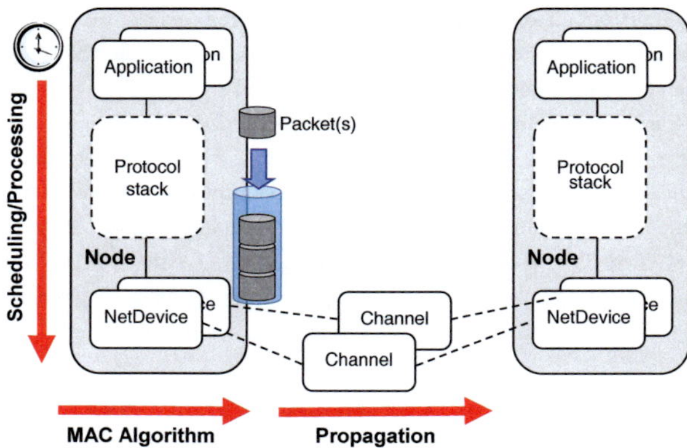

Figure 7.16: ns-3 node object overview and delays caused by scheduling/processing, MAC algorithm and propagation.

and hardware/software settings, from testbed measurements at the university lab. Thereby the introduced processing delay is estimated by the measurement from setups of up to five MSTAs which are placed to form a linear chain topology. In this arrangement it will be possible to measure the delay which is added by each hop in a wireless multihop scenario.

From these measurements a processing delay can be derived which will be included in the network simulations which finally leads to more accurate results even in complex scenarios.

Table 7.4 shows the protocol header lengths and payload of the packets for the ns-3 simulation and testbed measurements.

Table 7.4: Packet compilation in ns-3 and testbed.

	ns-3	testbed
Protocol	**Length** (Bytes)	**Length** (Bytes)
Radiotap Header v0	not used	38
IEEE 802.11 QoS Data	42	54
Logical-Link Control	8	8
IPv4	20	20
ICMP	8	8
Payload	222	172
Packetsize	300	300

The payload in the ns-3 simulation has been adjusted to meet the packet size of the testbed measurements. The differences arise from the radiotap header which is not used in the ns-3 simulation and the Mesh Address Extension subfield of the Mesh Control Field which is present in the measurements, however not implemented in ns-3. In this section the processing delay which is added at each hop in a linear chain topology is estimated based on the measurement of the round-trip-time in this testbed. Four scenarios have been measured with 2, 3, 4 and 5 MSTAs respectively. After each measurement run one MSTA is added at the end of the chain.

7.3.1 Testbed Setup

The measurement setup is based on up to five of the WLAN modules described in Section 7.1.2, which have been placed in the university lab. The open-source GNU/Linux distribution for embedded devices OpenWrt [16] is installed on each of the modules with the Version Chaos Calmer (SVN release 45514). Each module is configured as Mesh Station with the same mesh ID and channel based on the WLAN-Stack mac80211. By default, each MSTA automatically attempts to create peer links with MSTAs with the same mesh ID and channel. In order to keep the scenario simple, IEEE 802.11a is set as hardware mode and only Chain0 of the WLAN module is used.

The modules are placed in the lab to model a linear chain topology (see Fig. 3.8). During the measurements it was ensured that each module has only an established peer link to its direct neighbors. It should be noted, that interference between adjacent neighbors could not be completely avoided due to the wireless propagation environment. The maximum possible datarate was adjusted to $6\,\mathrm{Mbits/s}$ (BPSK with coderate 1/2) on each of the modules. The round-trip-time for each of the scenarios is measured by performing a ping from the first to the last WLAN module of the linear chain topology. The ping was repeated 60.000 times for each configuration and was performed with an interval of $20\,\mathrm{ms}$ and $172\,\mathrm{bytes}$ of payload. As discussed, Tab. 7.4 shows the protocol and header length of the packet.

7.3.2 Measurement Results

Figure 7.18 shows the histogram of the round-trip-time (RTT) in ms of the setups with 2, 3, 4 and 5 MSTAs. The measured RTTs are recorded with a resolution of $10\,\mu\mathrm{s}$. For 2 MSTAs a dedicated peak of the round-trip-time can be seen.

This can be explained as shown in Fig. 7.17. The processing delay is higher than $\mathrm{SIFS} + t_{\mathrm{ack}} + \mathrm{AIFS}$ and the MSTA achieves immediate access to the medium because the MSTA determines that the medium is idle for a time period greater than or equal to AIFS. Backoff, which is shown in Fig. 7.17, thus is not calculated[3]. The duration of the packets can be calculated by Eq. (3-27) as shown in Section 3.2.

---

[3]This result will also be used for interpretation of the results in Chapter 8.

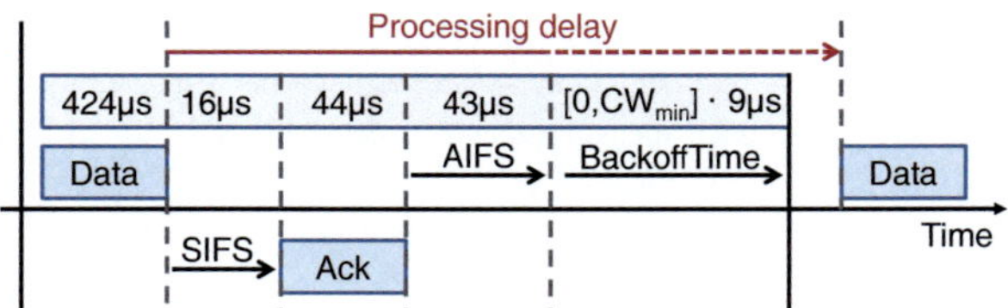

Figure 7.17: One hop duration with processing delay.

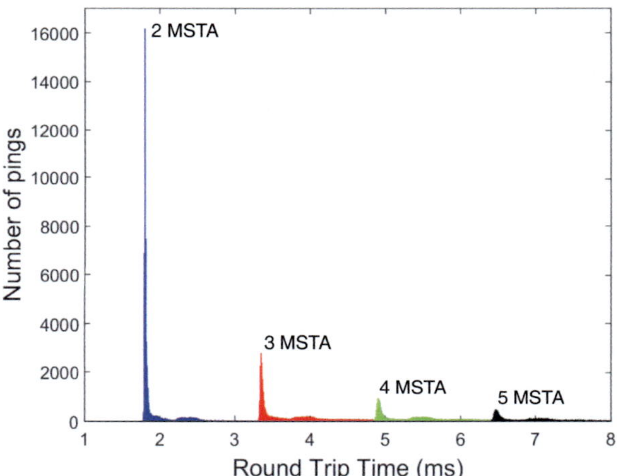

Figure 7.18: Histogram of round-trip-times for the measurement setups of chains with 2, 3, 4 and 5 MSTAs.

The Jitter of the round-trip-times is a result of the scheduling mechanism of the kernel. The wireless network interface device (WNIC) is implemented as SoftMAC. This means that the MAC-layer is essentially implemented in software. Only time critical parts such as the distributed coordination function and the Rx/Tx-State Machine are implemented in hardware.

In order to compare these results with those obtained by the ns-3 simulation as a first step a single processing delay based on the measurement results will be calculated as described below. The ICMP echo request or reply are generated at the first or last MSTA respectively. In the intermediate nodes only the MAC and PHY-layer are involved by forwarding the particular messages. In order to estimate a processing delay per hop a differentiation between end nodes and intermediate nodes is needed. The modules rely on the same hardware/software configuration and workload. Therefore it is assumed that the software execution time is the same in each of the modules, however, dependent on the actual state of the CPU because of time sharing with other tasks. The additional processing amount which occurs by the generation of the ICMP echo request and reply messages has to be considered only at the end nodes.

The round-trip-time is composed of the packet duration t_p, the processing delays of the MAC and PHY t_pd at each node and the processing delay of the IP and ICMP processing t_pd2 at the end nodes. It should be noted, that it is not possible to distinguish between the processing caused by the preparation of the echo request or reply message within the performed measurements. Therefore these processing delays will be equally weighted.

The following calculation of the processing delays is based on the schematic shown in Fig. 7.19.

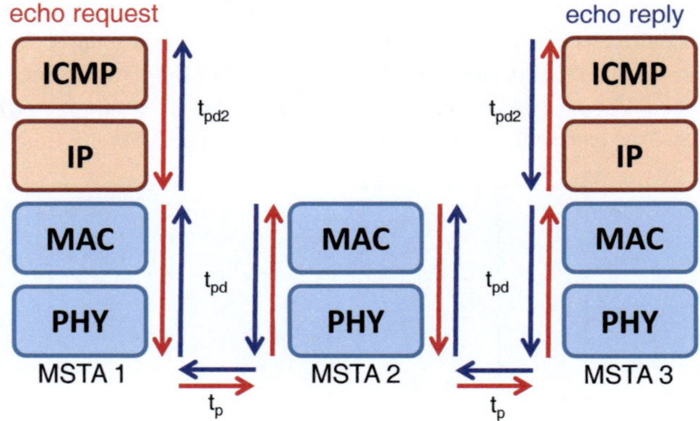

Figure 7.19: Breakdown of processing delays of PHY, MAC and IP/ICMP layer.

The calculation of the round-trip-time for one hop and for n hops can be derived from Fig. 7.19 and is shown in Eq. (7-5).

$$\begin{aligned} 1\,\mathrm{Hop} : \mathrm{RTT}_1 &= 2t_p + 4t_{pd} + 4t_{pd2} \\ n\,\mathrm{Hops} : \mathrm{RTT}_n &= 2n \cdot t_p + 4n \cdot t_{pd} + 4t_{pd2} \end{aligned} \tag{7-5}$$

From the measurement results in Fig. 7.18 a single round-trip-time without statistical variance can be extracted for one hop $\mathrm{RTT}_1 = 1.8\,\mathrm{ms}$ and for two hops $\mathrm{RTT}_2 = 3.35\,\mathrm{ms}$ as a first estimate when sticking to the first peaks of the respective histograms. By using Eq. (7-5) and solving for t_pd and t_pd2 with $t_\mathrm{p} = 424\,\mu\mathrm{s}$, the processing delay t_pd can be estimated to be $175.5\,\mu\mathrm{s}$ and the processing delay t_pd2 can be estimated to be $62.5\,\mu\mathrm{s}$. This is also valid for the results of setups with 4 and 5 MSTAs.

7.3.3 Network Simulation Validation

The accuracy of the estimated processing delay $t_{pd} = 175.5\,\mu s$ can be validated by including a single processing delay without statistical variance in the ns-3 simulation. To this end the packet forwarding is scheduled for a dedicated time in the future in order to emulate the processing delay.

The delay thereby has been included in the ForwardDown member function of the MeshWifi-InterfaceMac class in order to delay the packets before sending them to the outgoing wifi-queue. Figure 7.20 shows the simulation results for the configuration with 2 and 3 MSTAs with a single processing delay of $2t_{pd} = 351\,\mu s$. Additionally, the offset $4t_{pd2} = 250\,\mu s$ has been included in the simulation results as an additional constant delay that is added once. Apart from the modeling of processing delay the setup for the simulation is equivalent to the one used in Section 3.2.2.1 with a PSR adjusted to 90%. The resulting amount of retransmission thereby leads to a histogram of the round trip times equivalent to the one in Fig. 3.13.

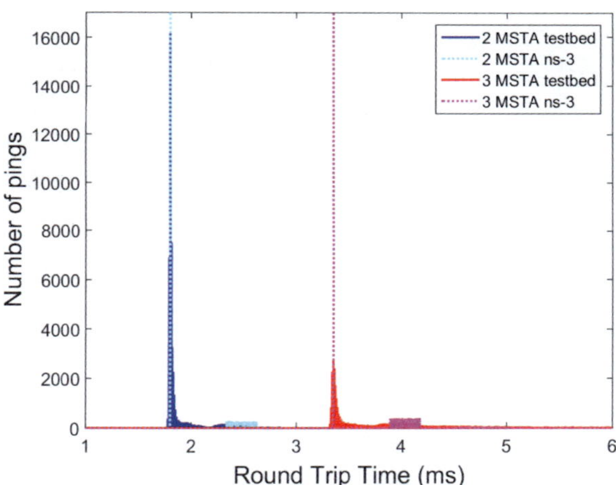

Figure 7.20: Histogram of round-trip-times for the measurement setups of chains with 2 and 3 MSTAs and corresponding ns-3 simulation with processing delay.

In principle the simulation result fits to the main peak of the measurement results. The RTT which can be associated with the first set of retransmissions is higher in the simulation than in the accomplished measurements. One reason for this is the inaccurate calculation of the ACKTimeout in the simulation as already described in Section 3.2.2.2. Another reason might be a non-standard conform implementation of the MAC of the Wi-Fi modules aiming on performance enhancement. Nevertheless, the delay of the first received retransmitted packets is slightly higher than expected by measurement data, which for the time being cannot be explained.

The difference is rather small and thus should not have an important impact on the simulation results for the overall DRS.

The performance of the scheduling and service execution of the single thread MIPS processor might be modeled by a M/G/1 processor sharing queue [127]. As a simplification in a first approach only the Poisson arrival process is modeled. The distribution of the RTTs that can be observed in the measurements is therefore modeled by a Poisson distribution with $\lambda = 2$. This value has been found by a heuristic approach. Thereby X is said to be Poisson (λ) distributed when

$$P_\lambda(X = i) = e^{-\lambda}\frac{(\lambda)^i}{i!} \quad (i \geq 0) \tag{7-6}$$

and $\lambda > 0$ is the parameter of the distribution [128]. For the generation of Poisson distributed random numbers in ns-3 a simple Poisson generator which is based on the multiplication of uniform random variables has been used (see Tab. 7.5) [128, p. 504].

Table 7.5: Poisson generator based upon the multiplication of uniform random variables. [128, p. 504]

```
X ← 0
Prod ← 1
WHILE True DO
    Generate a uniform [0,1] random variable U.
    Prod ← Prod · U
    IF Prod > e^−λ
        THEN X ← X + 1
    ELSE RETURN X
```

For the calculation of the processing delay the return value X is multiplied by $5\,\mu s$ and a mean value of $340\,\mu s$ is added in order to achieve an appropriate processing delay distribution. This delay has been included again in the ForwardDown member function of the MeshWifiInterfaceMac class in order to delay the packets before sending them to the outgoing wifi-queue.

The simulation result compared to the measurements of the configuration with 2 MSTAs is shown in Fig. 7.21. The simulation results are thereby plotted with the same resolution of $10\,\mu s$ as obtained by the recorded measurements. The Poisson distribution models the first part of the distribution of the round-trip-times, however, the exponentially decaying profile is not considered. The spreading of RTTs of the set of the first retransmission is similar to the measurements apart from the shift in delay as already discussed.

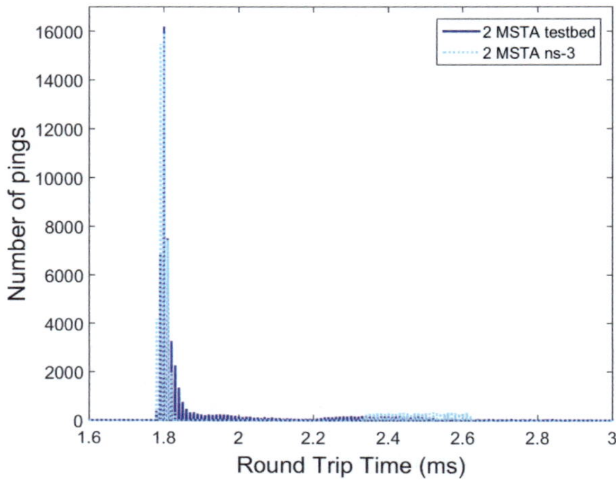

Figure 7.21: Histogram of round-trip-times for the measurement setups of chains with 2 MSTAs
and corresponding ns-3 simulation with processing delay and Poisson distribution.

7.3.4 Network Simulation of VoIP Communication Links with Processing Delay

Based on the results presented in the previous section a VoIP communication link will be
simulated in the linear chain topology. The simulation results with the introduced processing
delay distribution will be compared with the ns-3 standard implementation in terms of the
end-to-end delay and packet loss.

Compared to the simulations with ICMP the VoIP simulation is characterized by two individual
data streams competing for the channel access. Due to the hidden node problem, which is
fundamental in the linear chain topology, collisions may occur resulting in additional delay
and packet loss. Additionally, simulations are carried out with a dedicated packet success rate
which is emulated in the simulator in such a way that each packet has a dedicated PSR at
each transmission on the physical layer. A retransmission of a packet has again the same PSR.
As there is no VoIP communication model in ns-3, the communication link is modeled as two
individual UDP data streams as described in Chapter 6.

7.3.4.1 Simulation setup of VoIP links in linear chain topology

The duration of the connection is 60 sec during which 3000 UDP packets are sent for both data
streams. For each number of hops the simulation was repeated 100 times with different seeds.

The relevant simulation parameters are shown in Tab. 7.6. The IEEE 802.11s standard does not define the way of closing peer links, however, the ns-3 mesh model implemented two ways of doing this [86] as already discussed before. Peer links can be closed due to beacon loss and due to packet loss. If one of the parameters exceeds the threshold the peer link shall be closed. However, the default values of the ns-3 model MaxBeaconLoss=2 and MaxPacketFailure=2 are way too stringent. A closed peer link in the linear chain topology leads to a considerable number of lost packets and significant delay because of the additional peer link establishment process which is again carried out after a closed peer link. Therefore these values have been increased and guarantee that a peer link in the chain is always be maintained established in these simulations (see Tab. 7.6). The UnicastPreqThreshold and UnicastDataThreshold has been adjusted to 10 each in order to ensure that all peer links could be established. In that way PREQs are sent as unicast and acknowledged by the receivers. The traffic overhead by the unicast of the PREQ messages are low because every MSTAs has only a maximum of two neighbor MSTAs in this scenario.

Table 7.6: Simulation parameters in ns-3 for the simulation of VoIP links in a linear chain topology.

802.11s Peer Link	
MaxBeaconLoss	200
MaxRetries	3
MaxPacketFailure	200
MaxNumberOfPeerLinks	32
802.11s HWMP Protocol	
HWMPmaxPREQretries	5
UnicastPreqThreshold	10
UnicastDataThreshold	10
Communication Link	
Connections	2 Nodes (bidirectional)
Tx rate	68.8 kbps payload
Packetization Interval	20 ms
Voice Payload in Bytes	172
Duration of Connection	60 sec
Transport Protocol	UDP

7.3.4.2 Simulation results

Figure 7.22 shows the result of the simulation in ns-3 for different configurations for an ascending number of up to 20 hops. For each configuration the median of the delay and the minimum/maximum error-bounds in ms of the two UDP data streams from the leftmost MSTA (01) to the rightmost MSTA (02) of the linear chain are shown. The results obtained with the processing delay distribution (dashed line) are compared with the standard setup in ns-3

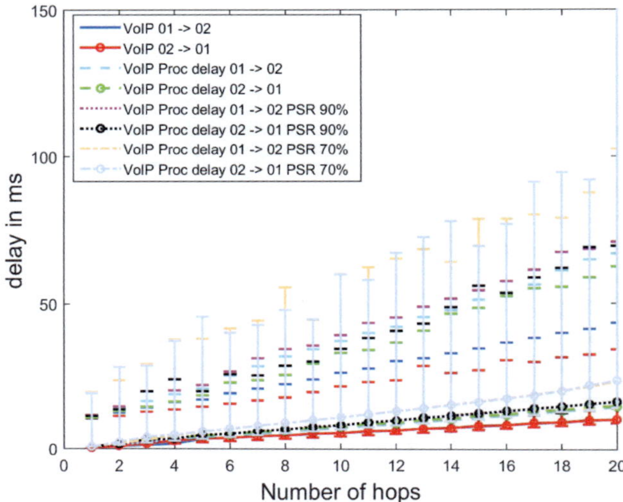

Figure 7.22: End-to-end delay of simulation setups realizing chains with 1 to 20 hops.

without processing delay (solid line). The simulation results with processing delay show the expected increase in delay by ascending number of hops. The dotted line shows the simulation result with a PSR of 90% and the dash-dotted line for a PSR of 70% respectively. A lower PSR results in an increased number of retransmissions due to the loss of packets and therefore leads to an increase in the end-to-end delay. The error-bounds indicate that even for a large number of hops, the QoS requirement of VoIP (end-to-end delay < 150 ms) is not violated for any of the packets. Only with the maximum number of 20 hops some packets exceed this constraint while the median value is still low.

Figure 7.23 shows the median packetloss and error-bounds of the 100 Monte Carlo simulation runs. Beside the configuration with PSR = 70% packetloss is almost zero. Even with a PSR of 70% the median packet loss is still below 1% for 20 hops, however, the error-bounds indicate that some simulation runs exceed this threshold. The presented results show that according to the constraints of delay and packet loss a transmission of up to 20 hops with sufficient QoS might be possible even for a PSR of 70%.

7.3.5 Conclusion

It has been shown that the processing delay occurring in the WLAN modules could be estimated in a forwarding scenario of a linear chain topology by measuring the round-trip-time and deriving a processing delay distribution. The network simulations have been validated by a

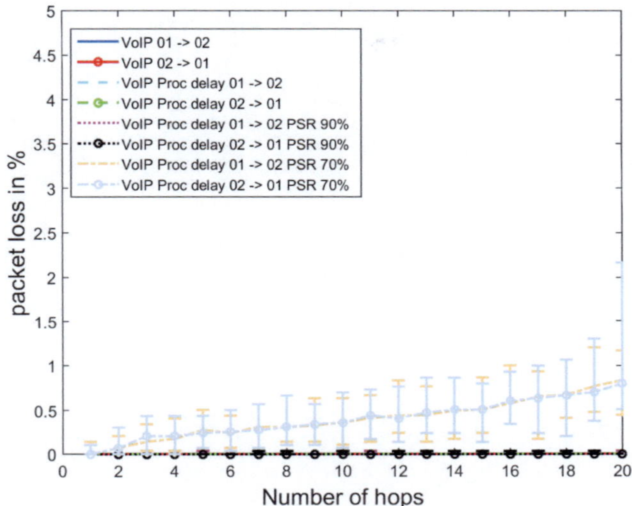

Figure 7.23: Packet loss of simulation setups realizing chains with 1 to 20 hops.

testbed setup in the university lab based on up to five IEEE 802.11a/g/n WLAN modules. The network simulator has been extended by a model for these processing delays and used for the performance estimation of VoIP communication links over multiple hops. It should be noted that the presented results show that according to the constraints of delay and packet loss a transmission of up to 20 hops with sufficient QoS might be possible even for a dedicated packet error rate of 30%. The presented model is valid only for the specified settings which correspond to the ones used in the VoIP analysis for the DRS in the following chapter. Different packet size and packet interval may result in different processing delay distributions which are also dependent on the used hardware.

Chapter **8**

System Performance of the DRS

Contents

So far network simulations of the proposed distributed sleep-wake algorithm introduced in Chapter 5 and validated in Chapter 6 have been carried out using the ns-3 Range-Propagation-Loss-Model (see also Sec. 3.2.3.1). Thereby the transmission range is set to a fixed value regardless of the transmission rate and packet size. This leads to results of the coverage and lifetime of the network comparable to those obtained with the system simulations using Matlab (in Chapter 5) when regarding modeling of the transmission on physical-layer, but taking into account the influence of the protocol stack from layer 2 upwards.

In this chapter the coverage, lifetime and network performance of the system will be analyzed by using an extended physical-layer model which allows for results which are expected to be much closer to those expected by setups using real hardware. To this end, the ITU Urban Micro cell (UMi) path loss model as described in Section 3.1.1 will be turned into a ns-3 model and will be used together with the new ns-3 error rate model described in Section 7.1.1. This error rate model shows a good correspondence with the measurement results of a typical WLAN module which has been shown in Sec. 7.1. The noise figure of the RF-Front-End of this WLAN module will also be taken into account (see Sec. 7.1.3). In order to get even more reliable simulation results, processing delays which are added in each MSTA due to the software execution time and scheduling mechanism of tasks in the CPU will be included in the simulations as well, using the model described in Section 7.3.

These final results including additional hardware parameters are assumed to give good indication on the feasibility and possible lifetime of the DRS with distributed sleep-wake algorithm by evaluating a VoIP connection of two individual rescuers using the whole protocol stack. Additionally, these results will indicate the performance of text messaging services when the stringent requirements of a VoIP connection are not fulfilled, but a connection through the network is still possible.

First, the achievable network performance will be analyzed at specific instants in time using the static network simulation methodology as introduced in Chapter 6. This means that the positions of the MSTAs which have been obtained from simulations of the distributed sleep-wake algorithm in Matlab are fixed and all MSTAs that have been taken from the result of the Matlab simulation remain active during the static network simulation. The static network simulation will be carried out for several specific time instances and the MSTAs correspond exactly to the MSTAs that have been found in the active state within the Matlab simulation which is a result from the executed algorithm. Second, the achievable network performance on an establishing network will be shown by a dynamic network simulation which has been introduced in Chapter 6 as well. Thereby the distributed sleep-wake algorithm is carried out in the network simulation itself. This dynamic network simulation includes also the influence of the protocols during start up of the disaster recovery network.

In order to differentiate between the influence of the processing delay model and the path loss model on network performance both are analyzed in different simulations in the following two sections.

All ns-3 simulations in the following sections have been carried out with the configuration of the sleep-wake algorithm with threshold for the number of neighbors $NoN_{Th} = 7$ which showed promising results throughout Chapter 6.

First, the results when using the processing delay model will be shown in Section 8.1. Second, the influence of the path loss model using the new error rate model is shown in Section 8.2.

8.1 Network Simulation of the DRS including Processing Delays

The static ns-3 simulations have been carried out with the configuration of the sleep-wake algorithm with threshold $NoN_{Th} = 7$ as well as for the initial configuration with full set of activated nodes. The results for the probability of successful calls which will be used in this section as baseline for the comparison with the simulation results carried out with processing delay model are originally shown in Section 6.1.2, Tab. 6.3. The simulation parameters also remain the same and can be taken from Tab. 6.2. The static ns-3 simulation is set up with the coordinates of the MSTAs as simulated in the coverage simulation in Matlab (Section 5.2.3) using Euclidean distance metric with 100 static MSTAs and two additional user MSTAs, where only nodes in the Node Active state at certain times T_0, T_1, T_2, T_3 and T_4 are used and remain active during the simulation. The simulations are carried out for the initial set up at time T_0=0 min and at times T_1=22 min, T_2=80 min, T_3=300 min and T_4=600 min. As stated in Section 6.1 T_1 corresponds to an instance in time where the settling process of the sleep-wake algorithm is still ongoing. At time T_2 the algorithm has settled and a stable thinned out network is expected. At time T_3 nodes previously active run out of power leading to a reduced overall coverage. At time T_4 the set of nodes is further reduced as most nodes are already out of power. T_0 is the start of the simulation with all MSTAs active.

In order to compare the results with those shown in the previous chapters, a fixed transmission radius of $r_0 = 60$ m has been chosen for the ns-3 Range-Propagation-Loss-Model (RPLM).

Figure 8.1 shows the probability of successful calls with processing delay model and compared to the simulation results of the static network simulations without processing delay model originally shown in Tab. 6.3. Additionally, Tab. 8.1 shows the detailed numbers of the results of possible calls as well as the mean delay of the communication link based on 1000 Monte Carlo simulation runs.

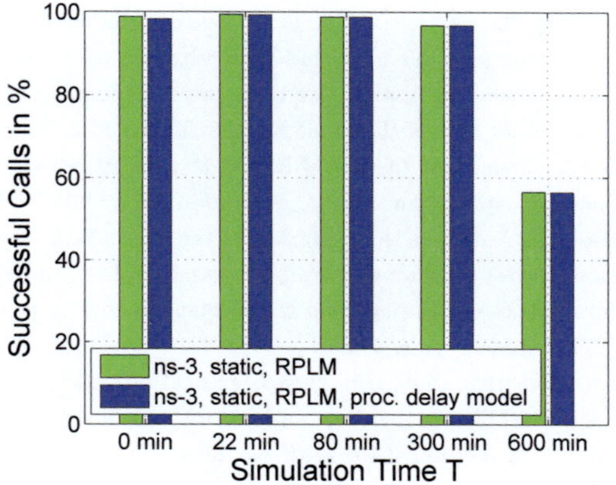

Figure 8.1: Probability of successful calls – static network simulation – comparison with processing delay model with $\text{NoN}_{Th} = 7$.

The results obtained using the processing delay model have a similar performance of successful calls when compared to those without processing delays. This leads to results for the probability of a successful call of 99% at times T_1 and T_2 and a probability of a successful call still higher than 96% at time T_3. At time T_4 the probability of a successful call is reduced to 56% which is a result of the reduced covered area of the remaining MSTAs. Thereby it should also

Table 8.1: Detailed numbers of the probability of successful calls and mean delay – static network simulations – comparison with processing delay model.

	Ratio of Successful Calls		mean delay in ms	
	RPLM	RPLM proc. delay	RPLM	RPLM proc. delay
NoN_{Th} = n.a., T_0=0 min	98.8%	98.3%	4.1 ms	5.4 ms
NoN_{Th} = 7, T_1=22 min	99.4%	99.3%	2.5 ms	3.8 ms
NoN_{Th} = 7, T_2=80 min	98.8%	98.8%	2.6 ms	3.8 ms
NoN_{Th} = 7, T_3=300 min	96.8%	96.8%	2.4 ms	3.6 ms
NoN_{Th} = 7, T_4=600 min	56.5%	56.4%	1.1 ms	1.6 ms

be noted that the simulated area is small and because of the used sleep-wake algorithm with number of neighbors metric the covered area is shrinked to the center of the area. MSTAs in the center of the area will still have connection to the network and allow VoIP calls. This has been also discussed in Section 5.2.3 where the covered area and lifetime of the system provided by the sleep-wake algorithm was analyzed by the help of a Matlab simulation at a considerably higher level of abstraction.

As stated previously, it should be noted, that the lifetime can be considerably higher when assuming a higher number of initially available MSTAs which would be a reasonable assumption [17]. In order to keep the resources required for network simulations within a reasonable range, the number of initially available nodes has been restricted as in the network simulations of Chapter 6 to 100, corresponding approximately to twice the number of nodes necessary to achieve $P_{\text{covered area}}$ = 90% for 90% of the whole area of $250\,\mathrm{m} \cdot 250\,\mathrm{m}$. It can also be seen from the results in Section 5.2.3 that the lifetime will tend to zero after about 14 hours with the used parameters of battery capacity and power consumption of each individual MSTA. For the scenario in this section the results at the beginning of the execution of the algorithm (T_0, T_1 and T_2) are of special interest because of the higher number of active MSTAs and expected higher amount of interference which may lead to a higher number of packet retransmissions and therefore to a higher value of packet delay.

Finally, it should be noted, that the static network simulation will not contribute additional results on the lifetime of the DRS, as the sleep-wake algorithm is not carried out in the network simulation itself. Results on lifetime of the DRS for the static network simulation will therefore not be different from those obtained in Chapter 5 and thus are not further investigated here.

The fact that the performance obtained when using the processing delay model in the network simulations is very similar to the performance obtained without using the processing delay model can be explained as follows: the simulation area of $A = 250\,\mathrm{m} \cdot 250\,\mathrm{m}$ being relatively small only a low number of hops is used by each communication link which leads to a low amount of additional delays due to processing time during forwarding of the packets. The obtained results of the successful calls, which are calculated by the evaluation criterion of packet loss, delay and jitter (Eq. (6-1) introduced in Section 6.1.2), are only caused from the first criterion, namely packet loss. The mean delay of the communication link as shown in Tab. 8.1 is only slightly higher when including the processing delays which does not affect fulfilment of the VoIP criteria and thus leads to similar overall results. Further, it should be noted that the mean delay is higher in the simulations with more active nodes at T_0. This is happening either due to a higher probability of interference and therefore a higher number of retransmissions of packets or due to a non-ideal path selection during link establishment by the Hybrid Wireless Mesh Protocol which is as well a result of stronger interference and may lead to more hops for the communication link. This is not further investigated here, as the distributed sleep-wake algorithm leads after its settling time to a reduced number of active MSTAs and therefore to a much weaker level of interference (being mainly due to management traffic in the considered scenario of a single VoIP call) and therefore to a lower mean packet delay.

The extra mean delay using the processing delay model with the reduced set of active MSTAs at times T_1, T_2 and T_3 is about 1.2 ms. In order to explain these results let us first investigate the reason for the mean delay that is present even without using the processing delay model. The mean delay without additional processing delay is a result of the packet duration t_p and the MAC procedure only. The total duration of one additional hop is in the range[1] of $[463, 598]\,\mu s$ assuming packet reception without retransmissions. This leads to the conclusion that a mean value of 5 hops is found in the communication links at T_1, T_2 and T_3. The higher mean delay when using the processing delay model is a result of the modeling by a Poisson distribution with minimum delay $340\,\mu s$ (see Sec. 7.3) where the processing delay is usually higher than the time needed for the MAC procedure[2] (see Fig. 7.17).

If the medium is idle, the MSTA achieves immediate access to the medium and the backoff procedure is not invoked when using the processing delay model. Therefore the additional mean delay is the difference between the processing delay and the time required for the MAC procedure with random backoff (see also Fig. 7.17). Because of that an exact value of the additional mean delay cannot be deduced, but it is dependent on the value of the Poisson distribution of the processing delay model and the backoff procedure of the MAC. Of course, in the presence of interference the delay can be even higher due to packet loss and retransmitting packets.

In order to estimate a maximum communication range the results obtained in Section 7.3 are reconsidered here. It has been shown in Section 7.3 that a transmission of up to 20 hops with sufficient QoS is possible even for a packet error rate of 30%. This means that each individual packet of the communication link of each hop has a probability of successful reception of 70% only. The reduced PSR thereby can be understood as a basic modeling of a constant interference noise level which is assumed for each packet and which leads to a certain amount of retransmissions. If a packet is retransmitted after a non-successful first transmission attempt, this packet is retransmitted again with a probability of successful reception of 70% only. As

---

[1] The total duration of one additional hop equals to t_p + SIFS + t_{ack} + AIFS + $[0, \text{aCWmin}] \cdot \text{aSlotTime} = [463, 598]\,\mu s$ with $t_p = 360\,\mu s$, SIFS $= 16\,\mu s$, $t_{ack} = 44\,\mu s$, AIFS $= 43\,\mu s$, aCWmin $= 15$ and aSlotTime $= 9\,\mu s$. See also Eq. (3-27) and Fig. 3.10.

[2] The MAC processing time equals to SIFS + t_{ack} + AIFS + $[0, \text{aCWmin}] \cdot \text{aSlotTime} = [103, 238]\,\mu s$.

the maximum delay allowed according to the analysis in Section 7.3 corresponds to approximately 20 hops and the delay is not considerably altered by processing delay as shown here, the communication range for VoIP connections is expected to be found in the range of approximately 1000 m based on the assumption of the used Range-Propagation-Loss-Model leading to a transmission range of $r_0 = 60$ m. If the number of hops of the VoIP communication link exceeds the possible communication range, still the forwarding of text messages is possible for the users in the disaster area.

Let us further look on the scalability of the performed network simulations. It has been mentioned that the simulation area is relatively small and the scalability of the Hybrid Wireless Mesh Protocol for larger areas was not investigated in the previous simulations. During path discovery the mesh network is flooded by Path Request (PREQ) messages which may lead to a considerable amount of traffic in the network if the number of MSTAs is high. However, for the use case considered in this work with a low number of concurrent VoIP streams the results are still sufficient when 100 MSTAs are assumed within the simulation area (see T_0 in Tab. 8.1). The usage of the distributed sleep-wake algorithm then quickly reduces the number of active MSTAs in the network and therefore also the amount of management traffic. By using the distributed sleep-wake algorithm with number of neighbors metric all MSTAs have a rather comparable number of neighbor MSTAs which leads to a similar probability of interference for each MSTA throughout the network. Therefore the interference in the thinned out network will not be dependent on the size of the network itself which insures scalability.
The performance of the network for a larger number of concurrent VoIP streams was, however, not investigated and needs to be further studied.

The results obtained so far have been achieved by using the ns-3 Range-Propagation-Loss-Model which leads to uniform transmission ranges. This model has been used in order to achieve comparable results of the network coverage to those obtained by the simulations in Matlab. Thereby a transmission of the MSTAs with a distance to another MSTA higher than 60 m is never possible. Furthermore, it should be noted that transmitted packets of MSTAs at a distance slightly higher than the specified transmission range do not contribute to the interference noise level of the corresponding MSTA which may lead to better results due to less interference.
This will be further analyzed in the next section by using the ITU Urban Micro cell (UMi) path loss model described in Section 3.1.1 together with the new ns-3 error rate model described in Section 7.1.1. Using the error rate model instead of a fixed range model may lead to a higher number of retransmitted packets due to more interference which may also lead to higher packet loss and delay and therefore to a lower number of possible hops of a VoIP communication link.

8.2 Network Simulation of the DRS with Additional Hardware Parameters and Path Loss Model

The ITU Urban Micro cell (UMi) NLOS path loss model as described in Section 3.1.1 has been turned into a ns-3 model and is used together with the new ns-3 error rate model as described in Section 7.1.1. This path loss model has been implemented as shown by the appropriate equation in Tab. 3.1. Additionally, the shadowing and multipath fading margin for 90% coverage as used in the simplified link budget calculation and shown in Tab. 3.2 have been implemented as a fixed value Z using Eq. (3-15). In order to achieve comparable results to those shown in the previous sections with a transmission range of 60 m using the Range-Propagation-Loss-Model (RPLM), the PSR has been adjusted to 90% for a VoIP packet of size 2040 bits at a distance of 60 m. This means that at a distance of 60 m a packet of this size is successfully received with a probability of 90% when considering a point-to-point link without interference. For 10% of transmissions the packet will be discarded as a result of the error rate model and a retransmission will be scheduled according to the IEEE 802.11 MAC. This retransmission has again a probability of successful packet transmission of 90%. This leads to a probability of successful transmission of the VoIP packet of 99% for the first retransmission and 99.9% for the second retransmission. In that way transmission ranges are achieved that are comparable to the transmission range of 60 m when using the Range-Propagation-Loss-Model. The default value of the retry limit according to IEEE 802.11 [35] and also used in this simulation is 5. This attribute indicates the maximum number of transmission attempts that are made before a failure condition is indicated.

The PSR of 90% for a packet of size 2040 bits has been achieved by adjusting the transmit power level TxPower to 16.45 dBm. The noise figure is set to 3.185 dB as shown in Tab. 7.1. The Rx and Tx antenna gains are set to 5 dBi as used in Tab. 4.1. Modulation is set to BPSK with coding rate 1/2. The packetsize of 2040 bits is a result of the packetsize of the VoIP packet of 250 bytes according to Tab. 6.1 and additional PHY preamble, OFDM PHY header, PSDU, tail bits, and pad bits as shown in Eq. (3-27). For a point-to-point link without interference the PSR shrinks to 32% at a transmission distance of 63 m and to 0.66% at 65 m. A transmission at higher distances than 65 m therefore will not be possible in this simulation scenario. Additional interference noise which is a result of simultaneous transmissions of other MSTAs in the network will lead to smaller transmission ranges or to be more specific to lower PSRs due to a lower SINR (see Eq. (3-31)).

First, the achievable network performance will be analyzed by static network simulations in Section 8.2.1.

Second, the achievable network performance on an establishing network will be analyzed by a dynamic network simulation in Section 8.2.2.

8.2.1 Static Network Simulation

The static ns-3 simulation is set up with the same coordinates of the MSTAs as used in the previous section, whereby nodes in the Node Active state at certain times T_0, T_1, T_2, T_3 and T_4 as resulting from the Matlab simulations with Euclidean distance metric are used and remain active during the simulation. Table 8.2 shows the simulation parameters. At the end of this section additional simulations will be carried out using also the processing delay model.

Table 8.2: Simulation parameters – ns-3 static network simulation with additional hardware parameters.

Simulation Environment	
Area A	$250\,\mathrm{m} \cdot 250\,\mathrm{m}$
Number of MSTA	2 User MSTA, max 100 active MSTAs
YansWifiPhy & YansWifiChannel	
Propagation Model	UMi NLOS path loss model
Propagation Frequency	$5.5\,\mathrm{GHz}$
Error Rate Model	new error rate model
TxGain	$5\,\mathrm{dBi}$
RxGain	$5\,\mathrm{dBi}$
TxPower	$16.45\,\mathrm{dBm}$
RxNoiseFigure	$3.185\,\mathrm{dB}$
802.11s Peer Link	
MaxBeaconLoss	20
MaxRetries	4
MaxPacketFailure	25
MaxNumberOfPeerLinks	32
802.11s HWMP Protocol	
Dot11MeshHWMPmaxPREQretries	5
UnicastPreqThreshold	1
UnicastDataThreshold	1
Communication Link	
Tx rate	$80\,\mathrm{kbps}$ ($68.8\,\mathrm{kbps}$ payload)
Packetization Interval	$20\,\mathrm{ms}$
Voice Payload in Bytes	172
Start Time	uniform $[5,35]\,\mathrm{sec}$
Stop Time	uniform $[155,185]\,\mathrm{sec}$
Transport Protocol	UDP

Figure 8.2 shows the probability of successful calls with UMi NLOS path loss model and new error rate model compared to the results of static network simulation with the ns-3 Range-Propagation-Loss-Model originally shown in Tab. 6.3 based on 1000 Monte Carlo simulation runs. Additionally, Tab. 8.3 shows the detailed numbers for Fig. 8.2.

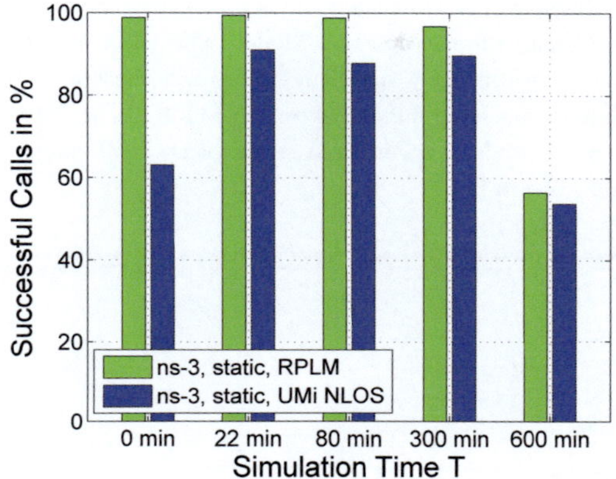

Figure 8.2: Probability of successful calls – static network simulation – comparison of Range-Propagation-Loss-Model (RPLM) and UMi NLOS path loss model with NoN_{Th} = 7.

Table 8.3: Detailed numbers of the probability of successful calls – static network simulation – comparison of Range-Propagation-Loss-Model (RPLM) and UMi NLOS path loss model with NoN_{Th} = 7.

	Ratio of Successful Calls	
	RPLM	UMi NLOS
NoN_{Th} = n.a., T_0=0 min	98.8%	63.1%
NoN_{Th} = 7, T_1=22 min	99.4%	91.0%
NoN_{Th} = 7, T_2=80 min	98.8%	87.8%
NoN_{Th} = 7, T_3=300 min	96.8%	89.7%
NoN_{Th} = 7, T_4=600 min	56.5%	53.8%

The results show a lower number of successful calls where a reduction of about 10% at T_1, T_2 and T_3 is found when compared to the results which have been obtained with the Range-Propagation-Loss-Model. This leads to a ratio of successful calls of 88% to 91% which would still be sufficient for the proposed DRS of this work if we stick to a required performance criterion of 90% successful calls.

However, for the initial configuration with all nodes active at T_0 a significant performance degradation of 35% can be observed. The reason is the higher probability of interference with all nodes active. When using the Range-Propagation-Loss-Model this behavior does not occur as the packet is either received at the transmit power level or when out of range with

$-1000\,$dBm (effectively zero). Therefore nodes out of range do not contribute to the interference noise when using the Range-Propagation-Loss-Model. This effect does not only occur during packet transmission. It is also likely that interference will cause a non-successful peer link establishment which leads to a lower number of connected MSTAs or will cause a non-successful path establishment which leads to a complete failure of the VoIP link. This can be further analyzed by the results shown in Fig. 8.3.

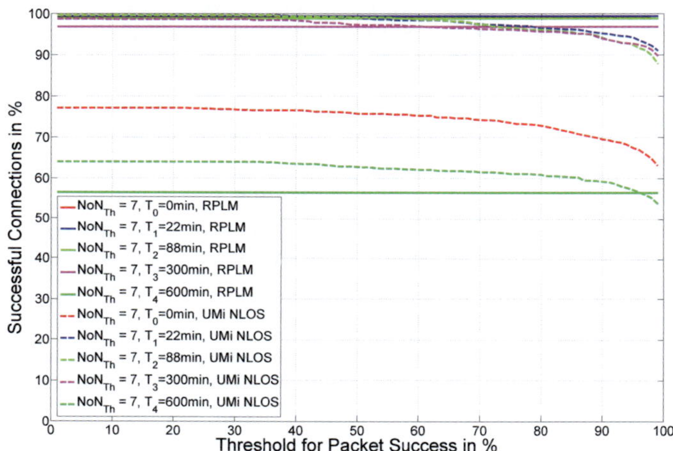

Figure 8.3: Probability of successful connections for different values for the thresholds of packet success from 1% to 99% – static network simulation – comparison of Range-Propagation-Loss-Model (RPLM) and the UMi NLOS path loss model with NoN$_{Th}$ = 7.

The obtained results for successful calls shown so far are calculated by the evaluation criterion of packet loss, delay and jitter as given by Eq. (6-1). The evaluation criterion of 1% packet loss for a VoIP stream is relatively stringent and the packet transmission of data packets for the forwarding of text messages may accept considerably higher values of packet loss for sufficient connections. This can be analyzed by varying the evaluation criterion of packet loss from 1% to 99% which is shown in Fig. 8.3. A low packet success threshold of only 1% also shows the number of links that have completely failed when compared to the previously obtained results with RPLM. The probability of a successful call (threshold for packet success = 99%) at T_0 (red dashed line) is 63.1%. For a packet success threshold of 1% the probability is increased to 78%. This shows that for about 20% of the calls a connection of the link was not established at all despite the higher number of possible connections due to the higher number of active MSTAs. This result also shows, that when using the distributed sleep-wake algorithm and reducing the number of active MSTAs, a higher probability of successful calls is achieved due to a lower amount of interference.

For text messaging services a even lower value of packet loss can be accepted than the stringent values of the VoIP connection. In contrast to VoIP, text messaging tolerates much higher latencies and is infrequent and asynchronous. With an appropriate application it is possible to store the messages when out of range and forward the messages later on. The packet size of text messages can be assumed similar to the simulated VoIP packets (compare e. g. the original character limit of a Twitter message which is 140 characters and which was an arbitrary choice based on the 160 character limit of the short message service (SMS)). With 160 seven-bit characters the data size of SMS thereby is 140 bytes which is slightly below the payload of 160 bytes of the simulated VoIP packets in this section. The packetization rate of the VoIP stream in this section was 50 pps. Referring this to the service of text messaging would mean that the two users are sending a text message every 20 ms to each other. This is of course not a realistic scenario for text messaging, but shows that a large number of text messages would be possible. For a low number of text messages a result for successful reception of the text messages close to the minimum of the threshold for packet success is possible. A statement on the performance of text messaging, however, is difficult based on these results. If we stick to a required packet success threshold of 50%, a successful connection can be still established for more than 98% of the situations at T_1 and T_2 and more than 96% of the situations at T_3 which shows that a large number of packets are received during this simulation.

The network capacity, however, should be analyzed in a different simulation approach which includes also a large number of concurrent message transmissions up to the saturation of the network.

The lifetime of the system can be analyzed by simulations on system level as described in the previous section. The probability of a successful call at time T_4 and using the UMi NLOS model is only slightly reduced to 53.8% when compared to the result obtained by the Range-Propagation-Loss-Model. The low successful call probability thereby is essentially a result of the reduced covered area of the remaining MSTAs as already shown for the RPLM. But it should also be noted, that with a lower packet success threshold ($<95\%$) a higher probability of a successful connection than with the Range-Propagation-Loss-Model can be achieved (see Fig. 8.3). An equivalent result is found from Fig. 8.3 for T_3 and PSR thresholds below 70%.

From these results it can be seen that with RPLM a constant rate of successful calls is achieved regardless of the specified value for the packet loss for all time instances. This means that a link is either established and has a low packet loss or is completely not working which is the result of the fixed range model RPLM. When using the UMi NLOS path loss model in combination with the new error rate model the probability of successful calls can be even higher at low packet success threshold values compared to RPLM. The reason is that all concurrent packets are treated with their exact power levels which are used for the SINR calculation. Previously by using the RPLM a packet was dropped if a concurrent transmission occurred because of the same power levels used within the transmission range of 60 m regardless of the distance. The adjustment of the PSR to 90% for a VoIP packet of size 2040 bits at a distance of 60 m that is achieved for a point-to-point link when using the setup as described for the UMi NLOS path loss model and in combination with the new error rate model allows for slightly higher transmission distances than 60 m, however with low SINR and therefore lower packet success rate. This explains also the decreasing percentage of successful calls at higher packet success threshold values to the lowest values at a threshold of 99% packet success for the connections shown in Fig. 8.3. If one of the single transmission links from the several hops has a low SINR value and therefore a certain packet loss, the hard limit of 99% packet success cannot be achieved.

This is a general limitation of wireless mesh networks. One link with low quality in the chain of the transmission over several hops will cause an overall low quality of the whole communication link.

Table 8.4 shows additional results with UMi NLOS path loss model, new error rate model and processing delay model as used in the previous section compared to the static network simulation results with the ns-3 Range-Propagation-Loss-Model based on 1000 Monte Carlo simulation runs. The results show that when additionally using the processing delay model a similar performance of successful calls is achieved when compared to the simulations without processing delays. These results confirm those from the previous section where it was shown that the mean delay of the communication link is only slightly higher when including the processing delays which does not affect fulfilment of the VoIP criteria and thus leads to similar overall results.

Table 8.4: Probability of successful calls – static network simulation – comparison of Range-Propagation-Loss-Model (RPLM), UMi NLOS path loss model and processing delay model with $\mathrm{NoN}_{Th} = 7$.

	Ratio of Successful Calls		
	RPLM	UMi NLOS	UMi NLOS & proc. delay model
NoN_{Th} = n.a., T_0=0 min	98.8%	63.1%	43.6%
NoN_{Th} = 7, T_1=22 min	99.4%	91.0%	90.3%
NoN_{Th} = 7, T_2=80 min	98.8%	87.8%	88.7%
NoN_{Th} = 7, T_3=300 min	96.8%	89.7%	89.0%
NoN_{Th} = 7, T_4=600 min	56.5%	53.8%	54.7%

The results at T_0, however, show an even worse performance compared to the previous results. This can be explained as a result of the higher interference level at T_0 and in combination with the modeling of the additional processing delay. The simulation without processing delay model is dependent on the MAC procedure only. When using the access category best effort (AC_BE) a maximum of 16 backoff slots are used as random delay for the medium access. In contrast when using the processing delay model a MSTA may achieve immediate access to the medium because of the higher delay compared to the delay of the MAC procedure which leads to the fact that the MSTA determines that the medium is idle for a time longer of than or equal to AIFS and the random backoff is not calculated (see Fig. 7.17). The used Poisson distribution, however, has a lower range of distributed delay slots than the backoff procedure for the first packet forwarding which leads to a higher probability of collisions for the first transmission attempt of each hop in the network simulation. This is a problem of the implementation of the processing delay distribution and might not cover real hardware performance which includes itself a higher degree of randomness. Therefore these results are not further investigated here, but should be analyzed by a more accurate implementation of the processing delay model.

8.2.2 Dynamic Network Simulation

The dynamic network simulation reflects the implementation of the distributed sleep-wake algorithm as described in Section 5.2.3 when running as part of the network simulation, i. e. all the network management messaging induced by the alternating sleep-wake states of the MSTAs are taken into account by execution of the IEEE 802.11 protocol stack. Dynamic network simulation thereby is carried out as detailed in Section 6.2 and is using the parameters shown in Tab. 6.4 with the settings of the initial random active time uniformly distributed in the interval $(5, 25)$ min and Node Active or Node Sleep timers uniformly distributed in the interval $(8, 24)$ min. The simulations have been carried out only for the configuration of the sleep-wake algorithm with threshold for the number of neighbors $\text{NoN}_{Th} = 7$ and based on 500 Monte Carlo simulation runs due to increased simulation times when using the error rate model.

Table 8.5 shows the probability of successful calls with path loss model and new error rate model compared to the simulation results with the ns-3 Range-Propagation-Loss-Model originally shown in Tab. 6.5 based on a reduced number of 500 Monte Carlo simulation runs in order to evaluate possible effects of the lower number of simulation runs. The results show a performance degradation of about 25% which is a degradation of 15% more than shown in the static simulations of the previous section based on the evaluation criterion of a maximum packet loss of 1% of the VoIP call.

Table 8.5: Probability of successful calls – dynamic network simulation – comparison of Range-Propagation-Loss-Model (RPLM) and UMi NLOS path loss model with $\text{NoN}_{Th} = 7$.

	allowed packet loss	Ratio of Successful Calls	
		RPLM	UMI NLOS
$\text{NoN}_{Th} = 7$, T_1=22 min	1%	94.2%	68%
$\text{NoN}_{Th} = 7$, T_2=80 min	1%	95.4%	69.2%
$\text{NoN}_{Th} = 7$, T_1=22 min	99%	98.5%	88.2%
$\text{NoN}_{Th} = 7$, T_2=80 min	99%	97.6%	88.8%

Let us first have a look on the differences to the static network simulation which allows an explanation of the results. It has to be emphasized that the static simulation in Sec. 8.2.1 is using the position of nodes obtained from simulations with sleep-wake algorithm carried out in Matlab with a fixed transmission range of 60 m. In order to achieve comparable results, the PSR has again been adjusted to 90% for a packet of size of 2040 bits at a distance of 60 m when considering a single point-to-point link without interference from other MSTAs. In this section the dynamic network simulation of the sleep-wake algorithm is carried out in the network simulation itself where the shut down of a MSTA is based on the number of neighbor MSTAs in its peer link list. It has already been shown in the previous section that when using the UMi NLOS path loss model and new error rate model higher transmission distances than 60 m can be achieved, however with lower SINR and therefore lower packet success rate.

One method to analyze the behavior of the distributed sleep-wake algorithm in ns-3 is to investigate the mean number of active MSTAs which is a result of the execution of the algorithm. The mean number of active MSTAs achieved from the Matlab implementation of the LEA distributed sleep-wake algorithm with number of neighbors metric has been shown in Chapter 5, Fig. 5.17. These results are reconsidered and further compared to the mean number of active MSTAs of the dynamic ns-3 simulation with both, the RPLM and the UMi NLOS path loss model in conjunction with the new error rate model for $\text{NoN}_{Th} = 7$. The corresponding results are shown in Fig. 8.4.

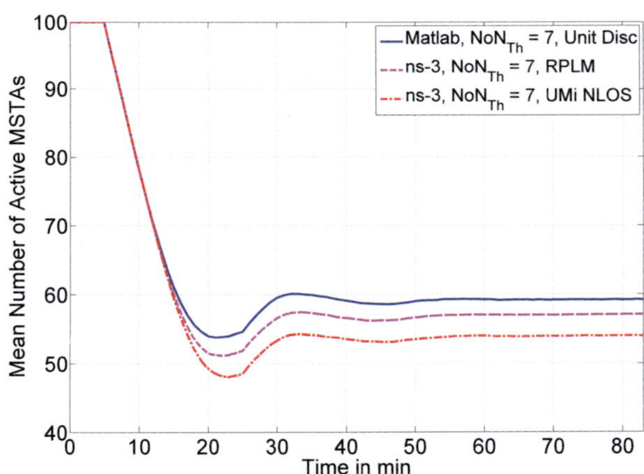

Figure 8.4: Mean number of active MSTAs according to the Matlab implementation of the LEA distributed sleep-wake algorithm with $\text{NoN}_{Th} = 7$ and as resulting from the dynamic ns-3 simulation with RPLM and UMi NLOS path loss model.

It has been already described in Section 6.2.2 that in the network simulation the two user MSTAs are part of the simulation itself and might be neighboring MSTAs of several relay MSTAs. This is in contrast to the Matlab simulation where the two user MSTAs are not part of the coverage simulation. This easily explains the result for the mean number of active MSTAs which is in the network simulation with RPLM about two MSTAs lower than in the Matlab simulation with unit disk model and without the two user MSTAs. The simulation with UMi NLOS path loss model (see the red dash-dotted line in Fig. 8.4) shows additionally a deviation of about 2 to 4 MSTAs after 20 min when compared to the simulation with RPLM. The lower number of active MSTAs is a result of the slightly higher transmission range when using the ITU path loss model which leads to a higher number of possible neighbor MSTAs and therefore a higher number of MSTAs which are shut down. However, the lower number of active MSTAs will also lead to a higher number of failed links or links with low quality which doesn't fulfil the VoIP criteria. A failed link thereby is assumed if a communication is not possible either because one of the user MSTAs is out of range of the connected cluster of nodes or because a link could not be established because of a non-successful path selection of the Hybrid Wireless Mesh Protocol. The mean number of active MSTAs as result of the executed sleep-wake algorithm which is shown in Fig. 8.4 shows also that the high interference at the beginning of the simulation with all MSTAs active does not affect the performance of the algorithm. The higher interference may have led to a number of non-successful peer links which would have led to a higher number of remaining active MSTAs according to the threshold condition NoN_{Th}. Figure 8.4 shows that this is not the case. In fact, the mean number of active MSTAs is lower in the simulation with UMi NLOS model when compared to RPLM.

Table 8.5 shows additional results for the probability of successful calls with a maximum packet loss threshold of 99% which includes all simulations where at least some packets could be delivered and a path was established. The results show that for more than 10% of the simulation runs a link completely fails which explains the first part of the degradation of the number of successful calls. In about 20% of the simulations a link was established and at least some packets could be delivered. However, the packet loss was too high to guarantee the QoS for VoIP. This can be seen in Fig. 8.5 which shows the probability of a successful transmission of the VoIP call for different thresholds of packet success from 1% to 99%. As stated earlier, if one of the links in the chain of MSTAs does have a connection with low PSR this will likely result in a non-successful call attempt.

Links with low quality can be a result of the following: the transmission range for a Mesh Peering Open frame which is used for the peer link establishment is higher than those for the data packets. This is simply given by the fact that the packet size of the Mesh Peering Open frame is 528 bits which leads to a PSR of 97.4% at a transmission range of 60 m when using the same physical-layer parameters as shown in Tab. 8.2 without interference. Additionally, the peering is retried several times before cancelling the peer link establishment procedure. This leads to peer links between MSTAs which are in a sufficient distance for establishing the peer link, but not for the transmission of data packets at relatively high PSR. This explains at least a major part of the result shown in Fig. 8.5. A certain number of MSTAs is shut down dependent on the chosen threshold NoN_{Th}. Because of the described conditions some of the remaining MSTAs will have peer links to other MSTAs with a low link quality. These links are causing a considerable performance degradation of successful VoIP calls.

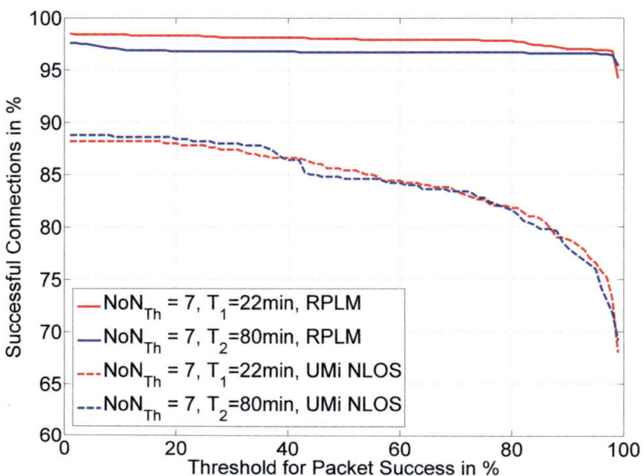

Figure 8.5: Probability of successful connections for different thresholds of packet loss from 1% to 99% – dynamic network simulation – comparison of Range-Propagation-Loss-Model (RPLM) and UMi NLOS path loss model with $\text{NoN}_{Th} = 7$.

Additionally, it should be noted that during the execution of the distributed sleep-wake algorithm links of shut down MSTAs can be closed due to a maximum beacon loss of successive beacons. This metric is used in the simulations to delete MSTAs from the peer link list which are shut down, but have not sent a peer link close message which is out of the scope of the current model. If this threshold for the loss of successive beacons is exceeded during runtime of active MSTAs a peer link would also be closed. However, because of the chosen value of 20 successive lost beacons it is very unlikely that in this set up a link with a bad connection in terms of a low PSR is closed due to this metric. The PSR of a beacon frame is similar to those of the Mesh Peering Open frame due to their comparable packet size. This means that a link with low quality stays always established despite its rather bad link conditions when transmitting data packets. Of course, if the threshold MaxPacketFailure is exceeded during data transmission, the peer link will also be closed. However, this would most likely lead to a lost connection of the VoIP stream because of the lack of alternative MSTAs and link path.

A solution to overcome this issue would be to specify a link quality parameter which allows to avoid links with low quality within a transmission chain by explicitly deleting these peer links from the peer link list. During the peer link establishment procedure the link quality to this end would have to be investigated and it will be decided whether the peer link is accepted or not. In order to avoid an unlimited attempt to set up peer links to MSTAs with low link quality, these MSTAs could be added to a "black list" whereby no more peer link establishment attempts are carried on for MSTAs on this list. In that way only peer links which satisfy a certain link quality are taken into account by the distributed sleep-wake algorithm.

Despite the described issues the results show that even though a VoIP connection was only successful in about 69% of the simulations, still in about 88% of the simulations packets could be delivered and a link could be established which still allows the forwarding of text messages for the users in the disaster area.
The number of successful calls can be even higher when avoiding links with low link quality conditions by the previously described mechanism.

8.2.3 Conclusion

For the use case considered in this work it could be shown that when referring to the static network simulation results a successful VoIP connection could be achieved with a probability of 90%. These results have been achieved by using an appropriate physical-layer and channel model, i.e. by using the Urban Micro cell (UMi) path loss model, additional hardware parameters and processing delays. The processing delay model itself showed a similar performance of successful calls when compared to those without processing delays. The lifetime of the system can be analyzed and improved by simulations on system level as shown in Chapter 5 without the need of simulation of the whole protocol stack. The performance degradation which could be observed at the beginning of the simulation is a result of the higher probability of interference with all nodes active. This result showed also that when using the distributed sleep-wake algorithm and reducing the number of active MSTAs, a higher probability of successful calls could be achieved due to a lower amount of interference. An impact of this interference on the performance of the distributed sleep-wake algorithm could not be observed when using the static network simulation.

The dynamic network simulation which refers to the implementation of the distributed sleep-wake algorithm in the network simulator itself and taking into account its interaction with the whole protocol stack shows a degradation of the number of successful calls of about 20% when compared to the static network simulation results. The performance degradation is related to links of a low quality within the transmission chain over several hops. The reason is the peer link establishment procedure which allows peer links with low link quality. To overcome this issue a mechanism was suggested which allows to avoid links with low quality to be added to the peer link list.
It should be noted that this can be considered to be a general problem of a multihop communication link because one link in the chain with a low quality will always degrade the overall performance of the link. Following this reasoning it is expected that results close to the static results can be achieved in the dynamic simulation when avoiding links with low quality.

This allows rescuers in the disaster area to set up a low number of VoIP calls within the network. The scalability to larger areas is not investigated in this work. However, it should be mentioned that the used metric which is based on the number of neighbor MSTAs leads to a similar amount of neighbor MSTAs throughout the network. This will also lead to similar amount of interference for a low number of VoIP calls. It is obvious that a large number of VoIP calls will likely overload the network. Text messaging services, however, tolerate much higher latencies, are infrequent and asynchronous and accept a even higher value of packet loss than the stringent values of the VoIP connection. In high load scenarios the transmission of text messages might therefore be still possible.

Finally, the capacity of the network is not investigated in this work and should be further analyzed for different scenarios of disaster recovery in future work.

Chapter 9

Conclusion and Outlook

Contents

9.1 Conclusion

A novel approach for quickly establishing a disaster recovery system, enabling the communication, notably between volunteer rescuers, during the first hours after occurrence of a disaster, has been proposed. The proposed DRS is realized by legacy, battery powered IEEE 802.11 wireless mesh capable devices, which are "left over" by their owners within the disaster region. It has been shown by Monte Carlo simulations including boundary effects that typical numbers of devices available in such disaster regions seem to be sufficient to provide network coverage of those regions to a reasonable extent. The demonstrated scalability of the system thereby plays an important role. An evaluation methodology for the coverage, namely the number of nodes which are part of the largest cluster, in percent of the total number, for achieving a certain covered area of the disaster area has been established. Thereby full connectivity of the network has not to be envisaged in a disaster region.

For system level simulations nodes are randomly placed in the disaster area. Different street topologies have been analyzed and compared to the random topology showing that a Manhattan topology with small block size representing the city structure of Mogadischu shows comparable results to the random topology. City structures with larger block sizes will lead to a larger amount of needed nodes for achieving a similar coverage when restricting the placed nodes located at the streets only. However, users will be located near the streets as well and the covered area is therefore restricted to these regions.

Key element of the newly proposed DRS is a distributed lifetime enhancement algorithm (LEA-DRS) that requires no central control unit and that allows to switch of devices that are not needed for realizing the coverage in a first time period, especially in order to save the remaining energy in the batteries for a later usage when the initially switched on devices run out of battery. It has been shown that with the proposed algorithm for thinning out the initially available network, the lifetime of a DRS can be considerably extended depending on the node density while keeping the required coverage of the disaster area. For the use case[1] considered lifetime could almost be doubled, which is key for bridging the time until professional rescuers arrive in the disaster region and will be able to bring up a dedicated network infrastructure. It should be emphasized that the lifetime has been analyzed by the moderate assumption of a low battery lifetime of about 4 to 6 hours of each MSTA, e. g. when using hand-held devices of users or battery powered routers only. MSTAs placed in cars which are located along side of streets will have a much longer lifetime because of the powerful batteries in these devices and the results obtained can be scaled to a larger overall lifetime.

As key parameter for the lifetime enhancement algorithm a threshold beyond which devices individually will decide to switch off has been established. Thereby the decision when a node changes its state is based on the number of neighbors of the corresponding node and without knowledge of the position of the nodes. It has been shown by Monte Carlo simulations of the sleep-wake algorithm that with a threshold of 7 or 8 for the number of neighbors a sufficient network coverage can be achieved.

The network performance of the wireless mesh network when using the distributed sleep-wake algorithm has been validated by ns-3 simulations of VoIP calls between individual rescuers. First, the network performance has been evaluated in a static setup at dedicated instants in time and thinned out network according to the proposed sleep-wake algorithm. After settling of the algorithm a probability of a successful call is in between 96% to 99.5% depending on the metric of the number of neighbor nodes. Second, the network has been evaluated in a dynamic setup in full use of the proposed algorithm. By including additional network traffic which is a result of the dynamic execution of the algorithm the successful call probability is slightly reduced depending on the configuration. For the configuration with 7 or 8 as threshold for the number of neighbor nodes the probability of a successful call is still higher than 94% which would be sufficient for the proposed DRS in this work.

The ns-3 mesh model in conjunction with the IEEE 802.11 MAC of the ns-3 wifi model has been analyzed in this work and solutions to overcome inaccuracies which are currently present have been proposed. The network simulations have been validated by a testbed setup in the university lab which was based on up to five IEEE 802.11a/g/n WLAN modules. A novel approach has been shown for estimation the processing delay occurring in the WLAN modules

---

[1]It should be noted that for the considered use cases the number of available MSTAs was always about the double of the number of MSTAs needed to achieve the required coverage of the considered area.

in a forwarding scenario of a linear chain topology by measuring the round-trip-time and deriving a processing delay distribution. The network simulator has been extended by a model for processing delays and used for the performance estimation of VoIP communication links over multiple hops. It should be noted, that the presented model is valid only for the specified settings. Different packet size and packet interval may result in different processing delay distributions which are also dependent on the used hardware.

Because of the inaccuracies in the existing ns-3 error rate models (YANS and NIST) for the calculation of the packet success rate and Signal-to-Interference-plus-Noise-Ratio a new error rate model for ns-3 has been developed in this work. Changes to both existing error rate models have been proposed as well, resulting in more reliable performance estimations. The ns-3 implementation has been compared with an exact physical-layer simulation using Matlab and measurements over an AWGN channel with a typical IEEE 802.11 a/b/g/n wireless LAN module. Thereby the new error rate model leads to results being comparable to real hardware performance.

Finally, the coverage, lifetime and network performance of the system has been analyzed by using an extended physical layer model which allows for results which are expected to be much closer to those expected by setups using real hardware. A degradation of the number of successful calls of about 20% has been shown for the dynamic network simulation when compared to the static network simulation results. The dynamic network simulation includes the implementation of the distributed sleep-wake algorithm in the network simulator itself and is taking into account its interaction with the whole protocol stack including the extended physical layer model. The performance degradation is related to links of low quality within the transmission chain over several hops which has to be further investigated. The reason is the peer link establishment procedure which allows for peer links with low link quality.

9.2 Outlook

This work focused on the feasibility, system design, hardware requirements and performance evaluation based on Monte Carlo and network simulations for the restoration of communication after the occurrence of a disaster. Future work should focus on implementability of the communication network on application level and to clarify the remaining issues.

The dynamic network simulation of the distributed sleep-wake algorithm showed a degradation of the number of successful calls of about 20% when compared to the static network simulation results. The performance degradation is related to links of low quality within the transmission chain over several hops. It was explained that this is a general problem of wireless multihop networks whereby the reason is the peer link establishment procedure which allows peer links with low link quality. To overcome this issue a mechanism is suggested which allows to avoid links with low quality to be added to the peer link list. This requires an extension to the MAC protocol.

To achieve a high performance under varying conditions, nodes need to adapt their transmission rate dynamically. The influence of rate control algorithms which are usually used in WLAN modules has not been analyzed in conjunction with the distributed sleep-wake algorithm developed in this work and should be further investigated. For all simulations a constant data rate of 6 Mbit/s has been used by Binary Phase-Shift Keying (BPSK) modulation and coding rate 1/2. This setting leads also to the highest transmission ranges due to the lower requirements on SNR. Adaptive rate control may lead to higher network capacities if some of the links are able to use higher order modulations. However, the issue of the peer link establishment procedure which allows peer links with low link quality remains and needs another solution.

The performance of the DRS has been evaluated throughout this work by VoIP streams between two rescuers. However, the performance of a number of concurrent VoIP streams has not been investigated in this work. The network capacity should be analyzed in a different simulation approach which includes also a large number of concurrent transmissions of different users. As fallback if the network conditions don't allow VoIP connections, text messaging services should be envisaged. Text messaging services tolerate much higher latencies, are infrequent and asynchronous and accept a even higher value of packetloss than the stringent values of the VoIP connection. Therefore the capacity of the network should be investigated in future work for different scenarios of disaster recovery by this type of application. As a further step traffic generated by possible applications on end-user equipment should be analyzed as well.

The simulations carried out in this work focused on the use case of a local mesh network. Internet connectivity by one or several gateways should also be envisaged and the performance of the MAC protocol and LEA should be further analyzed.

It should be noted that the LEA might be only useful at static MSTAs, e.g. in cars or smart home devices, whereby the battery consumption of devices carried by users is on their own responsibility. A user wants to decide itself when his device is active or not. A large number of active users may influence the performance of LEA if users are going to activate and deactivate their devices which should be further investigated. Thereby an adjustment of the LEA timing parameters might be needed.

The scalability to larger areas in terms of the network performance of the distributed lifetime enhancement algorithm was not investigated in this work. However, it is stated that the used metric for the shut down of MSTAs in LEA is based on the number of neighbor MSTAs which leads to a similar amount of neighbor MSTAs throughout the network. This will also lead to similar amount of interference for each MSTA of management traffic in low load scenarios. Therefore simulations for large areas which are cumbersome in view of simulation times, seem not to be required.

The security of the network, e.g. avoiding unauthorized access or establishment of the network without the occurrence of a disaster, was out of the scope of this work and should be further investigated. The functionality of a DRS mode based on wireless mesh networks in cars thereby might be regulated by law similarly to the in-vehicle emergency call (e-Call).

Finally, it should be noted that the findings of this thesis can also be used for other types of applications. WMNs are decentralized, easy to deploy, and characterized by dynamic self-organization, self-configuration, and self-healing properties. A number of MSTAs, e. g. for a sensor network, can be distributed in a factory hall without the need of network planing. Non-needed MSTAs are shut down according to the LEA sleep-wake algorithm and can be removed later on when not needed.

Appendix A

IEEE 802.11 MAC

Figure A.1 shows the basic access method, interframe spaces (IFS) and backoff procedure with random contention window (CW) size according to [35, Fig. 10-4].

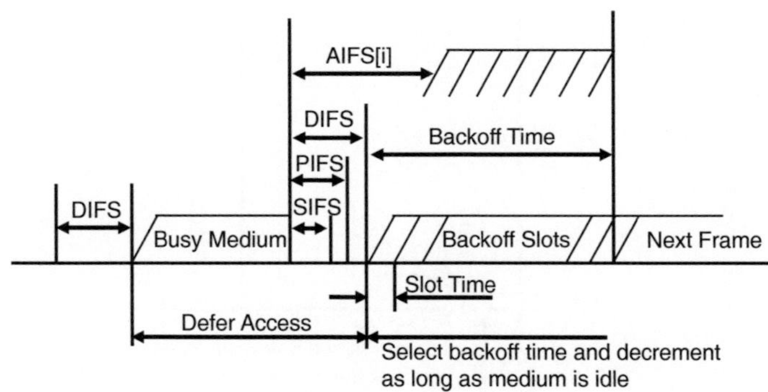

Figure A.1: Basic access method and IFS relationships according to [35, Fig. 10-4].

The basic IEEE 802.11 MAC protocol is the Distributed Coordination Function (DCF) which is based on carrier sense multiple access (CSMA). Thereby a station determines individually when to access the medium. Additionally, the DCF applies a collision avoidance (CA) mechanism, where stations perform a so-called backoff procedure before initiating a transmission. The minimum time the medium has to be idle before initiating a transmission is called DCF interframe space (DIFS) ($34\,\mu s$ in IEEE 802.11a [35, Eq. (10-6)]). After this time the stations keep sensing the medium for an additional random time called backoff time. A station initiates its transmission only if the medium remains idle for this additional random time. The duration of this random time is determined by each station individually, as a multiple of a slot time ($9\,\mu s$ according to the OFDM PHY specification with 20 MHz channel spacing [35, Tab. 17-21] as used in this work). It should be noted that if the medium has been idle for at least one DIFS, a frame might be immediately delivered without performing the backoff procedure. The random backoff time is calculated according to [35, Eq. (10-1)].

$$\mathrm{BackoffTime} = \mathrm{Random}() \cdot \mathrm{aSlotTime} \tag{A-1}$$

A random integer is drawn from a uniform distribution over the interval [0,CW], where CW is an integer within the range of values of the PHY characteristics between aCWmin and aCWmax (see Tab. A.1) [35, Section 10.3.3]. For legacy DCF with the values of aCWmin=15 and aCWmax=1023, the backoff time for the first retransmission is in the interval $[0, 31] \cdot \mathrm{aSlotTime}$. After each retransmission the CW value is increased by sequentially ascending integer powers of 2 minus 1. See Fig. A.2 for the example of the exponential increase of CW according to [35, Fig. 10-13].

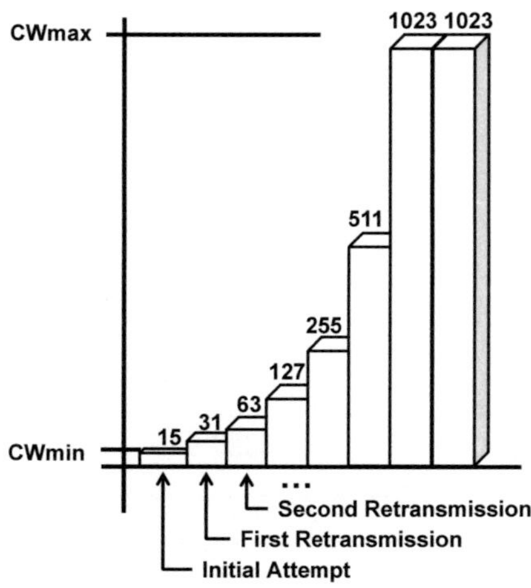

Figure A.2: Example of exponential increase of CW according to [35, Fig. 10-13].

Using these values it is less likely that collisions occur because of the reduced probability of repeated collisions by a larger CW sizes. Thereby all stations use the same value for CWmin, but select their random backoff time individually. Because of the same CWmin value of legacy DCF all stations have the same medium access priority in the DCF with no QoS support.

It should be noted that the short interframe space (SIFS) as shown in Fig. A.1 which is used e. g. for acknowledge notifications (ACK) gives them the highest priority access to the wireless medium because of the shortest interframe space. Thus no other stations can interrupt ongoing frame exchanges by initiating another transmission.

QoS support has been introduced in IEEE 802.11e by the hybrid coordination function (HCF). The mandatory Enhanced Distributed Channel Access (EDCA) mechanism is also used by the Mesh Coordination Function (MCF) which is present in MSTAs and used throughout this work. EDCA differentiates between four access categories (ACs) namely AC_BK for background traffic, AC_BE for Best Effort traffic, AC_VI for Video traffic and AC_VO for Voice traffic. Instead of DIFS the arbitration interframe space (AIFS[AC]) is used in order to allow a priorization of the different access categories. The AIFS[AC] has a minimum duration of DIFS, and can be enlarged by the AC with the help of the arbitration interframe space number (AIFSn[AC]). The AIFSn[AC] defines the duration of AIFS[AC] according to:

$$\text{AIFS[AC]} = \text{SIFS} + \text{AIFSn[AC]} \cdot \text{aSlotTime}, \quad \text{AIFSn[AC]} \geq 2 \qquad \text{(A-2)}$$

The values of aCWmin, aCWmax and AIFSn are shown in Tab. A.1. Further information can be found in [35].

Table A.1: Default EDCA parameters for each access category according to [35, Tab. 9-137].

AC	CWmin	CWmax	AIFSn
Background (AC_BK)	15	1023	7
Best Effort (AC_BE)	15	1023	3
Video (AC_VI)	7	15	2
Voice (AC_VO)	3	7	2
Legacy DCF	15	1023	2

Appendix B

Overview of ns-3 Mesh Module

This section shows an overview of changes of the ns-3 mesh model carried out during this work. Changes described in [23] include:

- Category codes and the categories compliant to IEEE 802.11-2012 Table 8-38–Category values.

- Information Elements (An adjustment of the element ID values was needed according to Table 8-54 of IEEE 802.11-2012).

- Mesh Peering Management element format changed according to IEEE 802.11-2012 Figure 8-370.

- Mesh Configuration element format changed according to IEEE 802.11-2012 Figure 8-363.

- PERR element format changed according to IEEE 802.11-2012 Figure 8-394.

With these changes the messages of the Peering Management Protocol and Hybrid Wireless Mesh Protocol will be transmitted compliant to IEEE 802.11-2016 and the resulting pcap trace files can be analyzed by Wireshark [83].

Supported features of the current mesh implementation are:

- Mesh Peering Management (MPM), including link close heuristics and beacon collision avoidance.

- Hybrid Wireless Mesh Protocol (HWMP), including proactive and reactive modes, unicast/broadcast propagation of management traffic, multi-radio extensions.

- 802.11e compatible airtime link metric [83].

Currently not supported features are:

- Mesh Coordinated Channel Access (MCCA).

- Internetworking: mesh access point and mesh portal.

- Security.

- Power saving.

- Path maintenance. (sending PREQ proactively before a path expires)

- Though multi-radio operation is supported, no channel assignment protocol is proposed for now (Correct channel switching is not implemented) [83].

Appendix C

Toroidal Distance Metric

A toroidal distance metric can be used to eliminate border effects (see also [64]).

Let $d\left(\begin{pmatrix} x_1 \\ y_1 \end{pmatrix}, \begin{pmatrix} x_2 \\ y_2 \end{pmatrix}\right)$ denote the usual Euclidean distance between two points $\begin{pmatrix} x_1 \\ y_1 \end{pmatrix}$ and $\begin{pmatrix} x_2 \\ y_2 \end{pmatrix}$ on a limited area $[0, x_{\max}][0, y_{\max}]$. The toroidal distance is then

$$d_{\mathrm{Tor}}\left(\begin{pmatrix} x_1 \\ y_1 \end{pmatrix}, \begin{pmatrix} x_2 \\ y_2 \end{pmatrix}\right) = \min\left\{ \mathrm{d}\left(\begin{pmatrix} x_1 \\ y_1 \end{pmatrix}, \begin{pmatrix} x_2 \\ y_2 \end{pmatrix}\right), \right.$$

$$d\left(\begin{pmatrix} x_1 + x_{\max} \\ y_1 \end{pmatrix}, \begin{pmatrix} x_2 \\ y_2 \end{pmatrix}\right), \mathrm{d}\left(\begin{pmatrix} x_1 - x_{\max} \\ y_1 \end{pmatrix}, \begin{pmatrix} x_2 \\ y_2 \end{pmatrix}\right),$$

$$d\left(\begin{pmatrix} x_1 \\ y_1 + y_{\max} \end{pmatrix}, \begin{pmatrix} x_2 \\ y_2 \end{pmatrix}\right), d\left(\begin{pmatrix} x_1 \\ y_1 - y_{\max} \end{pmatrix}, \begin{pmatrix} x_2 \\ y_2 \end{pmatrix}\right),$$

$$d\left(\begin{pmatrix} x_1 + x_{\max} \\ y_1 + y_{\max} \end{pmatrix}, \begin{pmatrix} x_2 \\ y_2 \end{pmatrix}\right), d\left(\begin{pmatrix} x_1 + x_{\max} \\ y_1 - y_{\max} \end{pmatrix}, \begin{pmatrix} x_2 \\ y_2 \end{pmatrix}\right),$$

$$d\left(\begin{pmatrix} x_1 - x_{\max} \\ y_1 + y_{\max} \end{pmatrix}, \begin{pmatrix} x_2 \\ y_2 \end{pmatrix}\right), \left. d\left(\begin{pmatrix} x_1 - x_{\max} \\ y_1 - y_{\max} \end{pmatrix}, \begin{pmatrix} x_2 \\ y_2 \end{pmatrix}\right)\right\}$$

$$\text{where } d_{\mathrm{Tor}}\left(\begin{pmatrix} x_1 \\ y_1 \end{pmatrix}, \begin{pmatrix} x_2 \\ y_2 \end{pmatrix}\right) \leq d\left(\begin{pmatrix} x_1 \\ y_1 \end{pmatrix}, \begin{pmatrix} x_2 \\ y_2 \end{pmatrix}\right).$$

Acronyms

DSSS Direct Sequence Spread Spectrum. 20

DYMO Dynamic Manet On Demand. 24

EDCA Enhanced Distributed Channel Access. 20, 54, 56, 75

EDCAF Enhanced Distributed Channel Access Function. 148, 149

EDCF Enhanced Distributed Coordination Function. 148

EIRP Equivalent Isotropically Radiated Power. 39

ERS Emergency- and Rescue Services. 12, 13, 79

ESS Extended Service Set. 20

ETSI European Telecommunications Standards Institute. 13

FEC Forward Error Correction. 71

FWA Fixed Wireless Access. 16

GSM Global System for Mobile communication. 13, 15

GTNetS Georgia Tech Network Simulator. 50

HR/DSSS High Rate DSSS. 20

HT/OFDM High Throughput OFDM. 20

HWMP Hybrid Wireless Mesh Protocol. 22–25, 55, 81, 136, 139, 145, 181, 182, 191

IBSS Independent BSS. 19

ICMP Internet Control Message Protocol. 56, 168

IEEE Institute of Electrical and Electronics Engineers. 4

ISM Industrial, Scientific and Medical Band. 29

ITU International Telecommunication Union. 16, 29, 31

LAN Local Area Network. 19, 20

LEA Lifetime Enhancement Algorithm. 7, 25, 43, 109, 112, 113, 217

LMR Land Mobile Radio. 12

LOS Line-of-Sight. 30

LTE Long Term Evolution. 13, 15, 16

MAC Medium Access Control. 10, 19, 20, 52, 59, 83, 134, 169, 183

MANET Mobile Ad-hoc Network. 15

MBSS Mesh Basic Service Set. 19, 81

MCCA MCF Controlled Channel Access. 20, 52

MCF Mesh Coordination Function. 20, 56

MPM Mesh Peering Management. 7, 18, 22, 117, 123, 129, 205

MSTA Mesh Station. 4–7, 9, 10, 22, 25, 31, 42, 55–57, 63, 80–84, 90, 104, 105, 107, 112, 113, 119, 120, 123, 129, 134, 139–141, 156, 158, 159, 165, 168, 178, 179, 182, 183, 186, 187, 190–192, 216

NIST National Institute of Standards and Technology. 54, 62, 65, 68–73, 146, 151

NLOS Non-Line-of-Sight. 29, 30, 38, 85

OFDM Orthogonal Frequency-Division Multiplexing. 20, 34, 57, 69–71, 73, 183

OLSR Optimized Link State Routing Protocol. 24, 25

PDF Probability Density Function. 33

PER Packet Error Rate. 62, 64, 216

PERR Path Error. 23

PHY Physical Layer. 10, 20, 21, 23, 34, 52, 57, 63, 66, 68, 83, 169, 183

PLCP Physical Layer Convergence Procedure. 57

PREP Path Reply. 23, 160

PREQ Path Request. 23, 55, 182

PSDU (PLCP) Service Data Unit. 57

PSR Packet Success Rate. 60, 62, 70, 146, 149, 154, 170, 183

QAM Quadrature Amplitude Modulation. 20, 66–69

QCU Queue Control Unit. 148

QoS Quality of Service. 20, 74, 75, 165

QPSK Quadrature Phase Shift Keying. 20

RMa Rural Macro. 30

RSSI Received Signal Strength Indicator. 111, 118, 120, 129, 150

RTT Round-Trip-Time. 57

SINR Signal-to-Interference-plus-Noise-Ratio. 41, 42, 61–64, 66, 69, 72, 197, 216

SNR Signal-to-Noise-Ratio. 69–73, 154

STA Station. 19, 20, 82, 83, 215

TETRA TErrestrial Trunked RAdio. 13

TIA Telecommunications Industry Association. 13

UHF Ultra-High Frequency. 12

UMi Urban Micro cell. 30, 31, 84, 85, 177, 182, 183, 193

UMTS Universal Mobile Telecommunications System. 13, 15

UPS Uninterruptible Power Supplies. 13, 14, 17

VHF Very High Frequency. 12

VoIP Voice over IP. 4, 6, 7, 9, 10, 16, 74–76, 80, 133, 134, 146, 165, 178, 183

WiMAX Worldwide Interoperability for Microwave Access. 13, 15, 16, 29

WLAN Wireless Local Area Network. 4, 5, 9, 15, 55, 83, 149

WMN Wireless Mesh Network. 4–6, 25, 27, 28, 39, 46, 49, 75, 84, 109

WSN Wireless Sensor Networks. 109, 110

YANS Yet Another Network Simulator. 50, 54, 62, 66, 68–73, 146, 151, 154

List of Figures

List of Tables

Bibliography

[1] D. Guha-Sapir, R. Below, and Ph. Hoyois. EM-DAT: The CRED/OFDA International Disaster Database - www.emdat.be. *Université Catholique de Louvain – Brussels – Belgium*, 2016.

[2] R. Miura, M. Inoue, Y. Owada, K. Takizawa, F. Ono, M. Suzuki, H. Tsuji, and K. Hamaguchi. Disaster-resilient wireless mesh network - experimental test-bed and demonstration. In *2013 16th International Symposium on Wireless Personal Multimedia Communications (WPMC)*, pages 1–4, June 2013.

[3] Y. Lien, H. Jang, and T. Tsai. A manet based emergency communication and information system for catastrophic natural disasters. In *2009 29th IEEE International Conference on Distributed Computing Systems Workshops*, pages 412–417, June 2009.

[4] T. Ngo, H. Nishiyama, N. Kato, Y. Shimizu, K. Mizuno, and T. Kumagai. On the throughput evaluation of wireless mesh network deployed in disaster areas. In *2013 International Conference on Computing, Networking and Communications (ICNC)*, pages 413–417, Jan 2013.

[5] Q. T. Minh, K. Nguyen, C. Borcea, and S. Yamada. On-the-fly establishment of multihop wireless access networks for disaster recovery. *IEEE Communications Magazine*, 52(10):60–66, October 2014.

[6] W. Lu, W. K. G. Seah, E. W. C. Peh, and Y. Ge. Communications support for disaster recovery operations using hybrid mobile ad-hoc networks. In *32nd IEEE Conference on Local Computer Networks (LCN 2007)*, pages 763–770, Oct 2007.

[7] D. Iland and E. M. Belding. Emergenet: robust, rapidly deployable cellular networks. *IEEE Communications Magazine*, 52(12):74–80, December 2014.

[8] K. Andersson and V. P. Kafle. Disaster-resilient mobile network architecture. In *2014 IEEE 11th Consumer Communications and Networking Conference (CCNC)*, pages 157–162, Jan 2014.

[9] Y. Shibata, N. Uchida, and N. Shiratori. Analysis of and proposal for a disaster information network from experience of the great east japan earthquake. *IEEE Communications Magazine - Lessons of the Great East Japan Earthquake*, 52(3):44–50, March 2014.

[10] NetHope. Information and Communication Technology Usage in the 2010 Pakistan Floods. *NetHope Case Study*, 2011.

[11] Google person finder. `https://google.org/personfinder/global/home.html`. Accessed: 2017-01-17.

[12] D. Goldman. Google gives '20%' to Japan crisis. *CNNMoney*, 2011.

[13] Federal Office of Civil Protection and Disaster Assistance – Bundesamt für Bevölkerungsschutz und Katastrophenhilfe (BBK). Guide for Emergency Preparedness and Correct Action in Emergency Situations. `http://www.bbk.bund.de/SharedDocs/Downloads/BBK/DE/Publikationen/Broschueren_Flyer/Fremdsprachliche_Broschueren_Ratg_u_fdNv/Disasters-alarm_Bro.pdf?__blob=publicationFile`, 2016.

[14] Federal Office of Civil Protection and Disaster Assistance – Bundesamt für Bevölkerungsschutz und Katastrophenhilfe (BBK). Press release – Smartphone as a lifesaver: An app for disaster situations. `http://www.bbk.bund.de/SharedDocs/Downloads/BBK/DE/Presse/Pressemeldung_2017/PM_smartphone_als_Lebensretter.pdf;jsessionid=9A104C155F5577E9CD887EA433E28FFD.2_cid330?__blob=publicationFile`, 2017.

[15] IEEE Computer Society. IEEE Std 802.11s. *Amendment 10: Mesh Networking*, Sept 2011.

[16] OpenWrt. `https://wiki.openwrt.org/`, 2016.

[17] K. Panitzek, I. Schweizer, T. Boenning, G. Seipel, and M. Muehlhaeuser. First responder communication in urban environments. *Int. J. Mob. Netw. Des. Innov.*, 4(2):109–118, August 2012.

[18] C. Hepner, S. Moll, and R. Muenzner. Influence of Processing Delays on the VoIP Performance for IEEE 802.11s Multihop Wireless Mesh Networks: Comparison of ns-3 Network Simulations with Hardware Measurements. In *Proceedings of the 9th EAI International Conference on Simulation Tools and Techniques*, SIMUTOOLS'16, pages 86–95, ICST, Brussels, Belgium, Belgium, 2016. ICST (Institute for Computer Sciences, Social-Informatics and Telecommunications Engineering).

[19] C. Hepner and R. Muenzner. Lifetime enhancement of disaster recovery systems based on IEEE 802.11s wireless mesh networks. In *11th IEEE International Conference on Wireless and Mobile Computing, Networking and Communications, WiMob 2015, Abu*

Dhabi, United Arab Emirates, October 19-21, 2015, pages 91–99. IEEE Computer Society, 2015.

[20] C. Hepner, R. Muenzner, and R. Weigel. Lifetime enhancement of disaster recovery systems based on IEEE 802.11s wireless mesh networks using a sleep-wake algorithm with minimum number of neighbors. In *2017 IEEE 13th International Conference on Wireless and Mobile Computing, Networking and Communications (WiMob)*, pages 218–226, Oct 2017.

[21] C. Hepner, A. Witt, and R. Muenzner. In depth analysis of the ns-3 physical layer abstraction for WLAN systems and evaluation of its influences on network simulation results. *BW-CAR Symposium on Information and Communication Systems (SInCom)*, 2015.

[22] C. Hepner, A. Witt, and R. Muenzner. Extended Abstract, A new ns-3 WLAN error rate model - Definition, validation of the ns-3 implementation and comparison to physical layer measurements with AWGN channel. Workshop on ns-3 WNS3 2015, Castelldefels (Barcelona), Spain. `https://www.nsnam.org/wp-content/uploads/2015/04/WNS3_2015_submission_34.pdf`, 2015.

[23] C. Hepner, A. Witt, and R. Muenzner. Extended Abstract, Validation of the ns-3 802.11s model and proposed changes compliant to IEEE 802.11-2012. Workshop on ns-3 WNS3 2015, Castelldefels (Barcelona), Spain. `https://www.nsnam.org/wp-content/uploads/2015/05/WNS3_2015_submission_33.pdf`, 2015.

[24] F. Ranghieri and M. Ishiwatari. Learning form Megadisasters – Lessons From The Great East Japan Earth-Quake. *International Bank for Reconstruction and Development, The World Bank*, 2014.

[25] M. Portmann and A. A. Pirzada. Wireless mesh networks for public safety and crisis management applications. *IEEE Internet Computing*, 12(1):18–25, Jan 2008.

[26] ETSI. TErrestrial Trunked RAdio (TETRA). `http://www.etsi.org/technologies-clusters/technologies/tetra`, 2017.

[27] Federal Office of Civil Protection and Disaster Assistance. Krisenmanagement bei einer großflaechigen Unterbrechung der Stromversorgung am Beispiel Baden-Wuerttemberg – crisis management in the case of a large interruption of the power supply using the example of Baden-Wuerttemberg. Case study `http://www.bbk.bund.de/SharedDocs/Downloads/BBK/DE/Publikationen/PublikationenKritis/Krisenhandbuch_Stromausfall_Kurzfassung_pdf.pdf?__blob=publicationFile`, 2011.

[28] Federal agency for public safety digital radio. `http://www.bdbos.bund.de`. Accessed: 2017-01-18.

[29] H. Nishizawa and T. Sakano. Overseas Deployment of MDRU: ITU Project in the Philippines, and MDRU Standardization Efforts. *Feature Articles: Movable and Deployable ICT Resource Unit—Architecture for Quickly Responding to Unexpected ICT Demand*, 2015.

[30] European Commission. eCall in all new cars from April 2018. `https://ec.europa.eu/digital-single-market/en/news/ecall-all-new-cars-april-2018`, 2015.

[31] B. Al-Bakri. Migration of eCall Transport. `https://www.ietf.org/proceedings/87/slides/slides-87-ecrit-1.pdf`, 2013.

[32] Open Garden. FireChat free messaging app. `https://www.opengarden.com/firechat.html`, 2017.

[33] P. Gardner-Stephen, R. Challans, J. Lakeman, A. Bettison, D. Gardner-Stephen, and M. Lloyd. The serval mesh: A platform for resilient communications in disaster & crisis. In *2013 IEEE Global Humanitarian Technology Conference (GHTC)*, pages 162–166, Oct 2013.

[34] P. Lieser, F. Alvarez, P. Gardner-Stephen, M. Hollick, and D. Boehnstedt. Architecture for responsive emergency communications networks. In *2017 IEEE Global Humanitarian Technology Conference (GHTC)*, Oct 2017.

[35] IEEE Computer Society. Part 11: Wireless LAN Medium Access Control (MAC) and Physical Layer (PHY) Specifications. *IEEE Std 802.11*, 2016.

[36] S. R. Das, C. E. Perkins, and E. M. Belding-Royer. Ad hoc On-Demand Distance Vector (AODV) Routing. RFC 3561, July 2003.

[37] M. Bahr. Proposed Routing for IEEE 802.11s WLAN Mesh Networks. In *Proceedings of the 2Nd Annual International Workshop on Wireless Internet*, WICON '06, New York, NY, USA, 2006. ACM.

[38] T. H. Clausen and P. Jacquet. Optimized Link State Routing Protocol (OLSR). RFC 3626, October 2003.

[39] J. Chroboczek. The Babel Routing Protocol. RFC 6126, April 2011.

[40] Open mesh development platform for b.a.t.m.a.n. (better approach to mobile ad-hoc networking). `https://www.open-mesh.org/projects/open-mesh/wiki`. Accessed: 2016-12-12.

[41] Research Institute on the Foundations of Computer Science. Babel – a loop-avoiding distance-vector routing protocol. `https://www.irif.fr/~jch/software/babel/`, 2017.

[42] Foerderverein Freie Netzwerke e.V. `https://freifunk.net/`, 2017.

[43] M. Piechowiak, P. Zwierzykowski, P. Owczarek, and M. Waslowicz. Comparative analysis of routing protocols for wireless mesh networks. In *2016 10th International Symposium on Communication Systems, Networks and Digital Signal Processing (CSNDSP)*, pages 1–5, July 2016.

[44] OMNeT++ Discrete Event Simulator. `https://omnetpp.org`, 2017.

[45] R. G. Garroppo, S. Giordano, and L. Tavanti. Experimental evaluation of two open source solutions for wireless mesh routing at layer two. In *IEEE 5th International Symposium on Wireless Pervasive Computing 2010*, pages 232–237, May 2010.

[46] open80211s. `https://github.com/o11s/open80211s/wiki/HOWTO`. Accessed: 2016-12-01.

[47] Y. Okumura, E. Ohmori, T. Kawano, and K. Fukuda. Field strength and its variability in VHF and UHF land-mobile radio service. *Review of the Electrical Communication Laboratory*, 16(9-10):825–873, Sep-Oct 1968.

[48] M. Hata. Empirical formula for propagation loss in land mobile radio services. *IEEE Transactions on Vehicular Technology*, 29(3):317–325, Aug 1980.

[49] E. Damosso and L. M. Correia. Digital mobile radio towards future generation systems. *COST 231 Final Report*, 1999.

[50] Electronic Communication Committee (ECC) within the European Conference of Postal and Telecommunications Administration (CEPT). The analysis of the coexistence of FWA cells in the 3.4 - 3.8 GHz band. *tech. rep. ECC Report 33*, May 2003.

[51] V. Erceg et al. Tgn channel models. `www.802wirelessworld.com`, May 2004.

[52] V. Erceg et al. Channel Models for Fixed Wireless Applications. *IEEE802.16.3c-01/29r4, Broadband Wireless Working Group, IEEE P802.16*, 2001.

[53] N. LaSorte, W. J. Barnes, B. Zigreng, and H. Refai. Performance Evaluation of a Deployed WiMAX System Operating in the 4.9GHz Public Safety Band. In *2009 6th IEEE Consumer Communications and Networking Conference*, pages 1–5, Jan 2009.

[54] International Telecommunications Union (ITU). Guidelines for evaluation of radio interface technologies for imt-advanced. *Report ITU-R M.2135*, 2008.

[55] P. Kyösti, J. Meinilä, L. Hentilä, X. Zhao, T. Jämsä, C. Schneider, M. Narandzic, M. Milojevic, A. Hong, J. Ylitalo, V.-M. Holappa, M. Alatossava, R. Bultitude, Y. deJong, and T. Rautiainen. IST-4-027756 WINNER II D1.1.1 V1.1 WINNER II interim channel models, Nov 2006.

[56] IEEE. Status of Project IEEE 802.11ax High Efficiency (HE) Wireless LAN Task Group. http://www.ieee802.org/11/Reports/tgax_update.htm, 2017.

[57] J. Kaushik and T. Rakesh. Submission. doc.: IEEE 11-14/0627r0. May 2014. Outdoor channel models for system level simulations. 11-14-0627-00-00ax-outdoor-channel-models-for-system-level-simulations.pptx, 2014.

[58] A. Molisch. *Wireless Communications*. Wiley-IEEE Press, 2005.

[59] T. S. Rappaport. *Wireless Communications Principles & Practice*. Prentice Hall Communications Engineering and Emerging Technologies Series, 2002.

[60] I. Tan, W. Tang, K. Laberteaux, and A. Bahai. Measurement and analysis of wireless channel impairments in dsrc vehicular communications. In *2008 IEEE International Conference on Communications*, pages 4882–4888, May 2008.

[61] R. Prasad. *OFDM for Wireless Communication Systems*. Artech House, 2004.

[62] ETSI. Harmonized European Standard – Broadband Radio Access Networks (BRAN); 5 GHz high performance RLAN; Harmonized EN covering the essential requirements of article 3.2 of the R&TTE Directive. *ETSI EN 301 893 V1.7.1*, 2012.

[63] G. Mao. *Connectivity of Communication Networks*. Springer, Cham, Switzerland, 2017.

[64] C. Bettstetter. On the minimum node degree and connectivity of a wireless multihop network. In *Proceedings of the 3rd ACM International Symposium on Mobile Ad Hoc Networking &Amp; Computing*, MobiHoc '02, pages 80–91, New York, NY, USA, 2002. ACM.

[65] N. A. C. Cressie. *Statistics for Spatial Data*. John Wiley & Sons, 1993.

[66] S. Mamechaoui, F. Didi, and G. Pujolle. A Survey on energy efficiency for wireless mesh network. *International Journal of Computer Networks and Communications (IJCNC) Vol.5, No.2*, Mar 2013.

[67] P. Gupta and P. R. Kumar. Critical power for asymptotic connectivity. In *Proceedings of the 37th IEEE Conference on Decision and Control (Cat. No.98CH36171)*, volume 1, pages 1106–1110 vol.1, Dec 1998.

[68] J. Deng, Y. S. Han, P. Chen, and P. K. Varshney. Optimal transmission range for wireless ad hoc networks based on energy efficiency. *IEEE Transactions on Communications*, 55(9):1772–1782, Sept 2007.

[69] J.-H. Chang and L. Tassiulas. Energy conserving routing in wireless ad-hoc networks. In *Proceedings IEEE INFOCOM 2000. Conference on Computer Communications. Nineteenth Annual Joint Conference of the IEEE Computer and Communications Societies (Cat. No.00CH37064)*, volume 1, pages 22–31 vol.1, March 2000.

[70] P. Chen, B. O'Dea, and E. Callaway. Energy efficient system design with optimum transmission range for wireless ad hoc networks. In *2002 IEEE International Conference on Communications. Conference Proceedings. ICC 2002 (Cat. No.02CH37333)*, volume 2, pages 945–952 vol.2, April 2002.

[71] L. Kleinrock and J. Silvester. Optimum transmission radii for packet radio networks or why six is a magic number. In *Conference Record, National Telecommunications Conference*, pages 4.3.2–4.3.5, Birmingham, A., December 1978.

[72] H. Takagi and L. Kleinrock. Optimal transmission ranges for randomly distributed packet radio terminals. *IEEE Transactions on Communications*, 32(3):246–257, Mar 1984.

[73] T.-C. Hou and V. Li. Transmission range control in multihop packet radio networks. *IEEE Transactions on Communications*, 34(1):38–44, Jan 1986.

[74] F. Xue and P.R. Kumar. The number of neighbors needed for connectivity of wireless networks. *Wireless Networks*, 10(2):169–181, 2004.

[75] L. Lazos and R. Poovendran. Stochastic coverage in heterogeneous sensor networks. *ACM Trans. Sen. Netw.*, 2(3):325–358, August 2006.

[76] Y. Wu, S. Fahmy, and N. B. Shroff. Energy efficient sleep/wake scheduling for multi-hop sensor networks: Non-convexity and approximation algorithm. In *IEEE INFOCOM 2007 - 26th IEEE International Conference on Computer Communications*, pages 1568–1576, May 2007.

[77] G. Xing, X. Wang, Y. Zhang, C. Lu, R. Pless, and C. Gill. Integrated coverage and connectivity configuration for energy conservation in sensor networks. *ACM Trans. Sen. Netw.*, 1(1):36–72, August 2005.

[78] A. de la Oliva, A. Banchs, and P. Serrano. Throughput and energy-aware routing for 802.11 based mesh networks. *Computer Communications 35*, 2012.

[79] P. Owczarek and P. Zwierzykowski. Review of simulators for wireless mesh networks. *Journal of Telecommunications and Information Technology*, 3:82–89, 2014.

[80] Omnet++. https://omnetpp.org/intro. Accessed: 2016-12-01.

[81] A. Khan, S. M. Bilal, and M. Othman. A performance comparison of network simulators for wireless networks. *CoRR*, abs/1307.4129, 2013.

[82] Wireshark project. http://www.wireshark.org/. Accessed: 2016-12-01.

[83] ns-3 project. https://www.nsnam.org. Accessed: 2016-12-01.

[84] ns-3 project. ns-3 Manual, Release ns-3.26, Oct 2016.

[85] T. Henderson and M. Lacage. ns-3 project annual meeting march 2013 - ns3 tutorial. www.nsnam.org/tutorials/consortium13/ns-3-tutorial-consortium.pdf, March 2013. Accessed: 2017-01-23.

[86] K. Andreev and P. Boyko. IEEE 802.11s Mesh Networking ns-3 Model. Workshop on ns-3 WNS3 2010, Malaga, Spain. http://www.nsnam.org/workshops/wns3-2010/dot11s.pdf, 2010.

[87] S. Papanastasiou, J. Mittag, E. G. Strom, and H. Hartenstein. Bridging the gap between physical layer emulation and network simulation. In *2010 IEEE Wireless Communication and Networking Conference*, pages 1–6, April 2010.

[88] M. Lacage and T. R. Henderson. Yet Another Network Simulator. *Proceeding from the 2006 workshop on ns-2: the IP network Simulator*, 2006.

[89] G. Pei and T. R. Henderson. Validation of OFDM error rate model in ns-3. *Boeing Research & Technology*, 2010.

[90] N. Baldo, M. Requena-Esteso, J. Nú nez Martínez, M. Portolès-Comeras, J. Nin-Guerrero, P. Dini, and J. Mangues-Bafalluy. Validation of the IEEE 802.11 MAC Model in the ns3 Simulator Using the EXTREME Testbed. In *Proceedings of the 3rd International ICST Conference on Simulation Tools and Techniques*, SIMUTools '10, pages 64:1–64:9, 2010.

[91] ns-3 bug reports. `https://www.nsnam.org/bugzilla`, 2016.

[92] WINLAB (Wireless Information Network Laboratory) - Network Simulator 3 (ns-3.16) Patches, 2016.

[93] IEEE Computer Society. Part 11: Wireless LAN Medium Access Control (MAC) and Physical Layer (PHY) Specifications. *IEEE Std 802.11*, 2012.

[94] A. Viterbi. Convolutional codes and their performance in communication systems. *IEEE Transactions on Communication Technology*, 19(5):751–772, October 1971.

[95] ns-3 project. ns-3 Model Library, Release ns-3.26, Oct 2016.

[96] M. Stoffers and G. Riley. Comparing the ns-3 propagation models. In *2012 IEEE 20th International Symposium on Modeling, Analysis and Simulation of Computer and Telecommunication Systems*, pages 61–67, Aug 2012.

[97] P. Fuxjaeger and S. Ruehrup. Validation of the ns-3 Interference Model for IEEE 802.11 Networks. In *2015 8th IFIP Wireless and Mobile Networking Conference (WMNC)*, pages 216–222, Oct 2015.

[98] J. G. Proakis. *Digital Communications*. McGraw-Hill, New York, 2001.

[99] P. Frenger, P. Orten, and T. Ottosson. Convolutional codes with optimum distance spectrum. *IEEE Communications Letters*, 3(11):317–319, Nov 1999.

[100] A. M. Camara and R. P. F. Hoefel. On the Performance of IEEE 802.11n: Analytical and Simulation Results. *XXIX Simposio Brasileiro de Telecomunicacoes - SBrt'11*, 2011.

[101] D. Haccoun and G. Begin. High-rate punctured convolutional codes for viterbi and sequential decoding. *IEEE Transactions on Communications*, 37(11):1113–1125, Nov 1989.

[102] A. Kashyap, S. Ganguly, S. R. Das, and S. Banerjee. Voip on wireless meshes: Models, algorithms and evaluation. In *IEEE INFOCOM 2007 - 26th IEEE International Conference on Computer Communications*, pages 2036–2044, May 2007.

[103] D. Niculescu, S. Ganguly, K. Kim, and R. Izmailov. Performance of voip in a 802.11 wireless mesh network. In *Proceedings IEEE INFOCOM 2006. 25TH IEEE International Conference on Computer Communications*, pages 1–11, April 2006.

[104] D. van Geyn, H. Hassanein, and M. S. El-Hennawey. Voice call quality using 802.11e on a wireless mesh network. In *2009 IEEE 34th Conference on Local Computer Networks*, pages 792–799, Oct 2009.

[105] T. Szigeti and C. Hattingh. *End-to-End QoS Network Design: Quality of Service in LANs, WANs, and VPNs*. Cisco Press, 2004.

[106] C. L. Huang and W. Liao. Throughput and delay performance of IEEE 802.11e enhanced distributed channel access (EDCA) under saturation condition. *IEEE Transactions on Wireless Communications*, 6(1):136–145, Jan 2007.

[107] H. Zhang, R. Zhao, and Z. Gao. Performance analysis of the 802.11 with a new model in nonsaturated conditions. In *Information Networking and Automation (ICINA), 2010 International Conference on*, volume 1, pages V1–344–V1–349, Oct 2010.

[108] K. Sanada, N. Komuro, and H. Sekiya. End-to-end throughput and delay analysis for IEEE 802.11 string topology multi-hop network using Markov-chain model. In *Personal, Indoor, and Mobile Radio Communications (PIMRC), 2015 IEEE 26th Annual International Symposium on*, pages 1697–1701, Aug 2015.

[109] G. Bianchi. Performance analysis of the IEEE 802.11 distributed coordination function. *IEEE Journal on Selected Areas in Communications*, 18(3):535–547, March 2000.

[110] D. Prasad and R. Goyal. Performance study of Voice in Wireless Mesh Network. *International Journal of Emerging Technology and Advanced Engineering*, 2013.

[111] A. Ksentini and O. Abassi. A comparison of voip performance over three routing protocols for ieee 802.11s-based wireless mesh networks (wlan mesh). In *Proceedings of the 6th ACM International Symposium on Mobility Management and Wireless Access*, MobiWac '08, pages 147–150, New York, NY, USA, 2008. ACM.

[112] M. Zogkou, A. Sgora, P. Chatzimisios, and D. D. Vergados. EDCA mechanism and mobility support evaluation in IEEE 802.11s WMNs. In *2014 6th International Congress on Ultra Modern Telecommunications and Control Systems and Workshops (ICUMT)*, pages 204–209, Oct 2014.

[113] S. Kristiansen, T. Plagemann, and V. Goebel. A methodology to model the execution of communication software for accurate network simulation. *ACM Trans. Model. Comput. Simul.*, 26(1):3:1–3:31, July 2015.

[114] A. Beifuss, D. Raumer, P. Emmerich, T. M. Runge, F. Wohlfart, B. E. Wolfinger, and G. Carle. A study of networking software induced latency. In *Networked Systems (NetSys), 2015 International Conference and Workshops on*, pages 1–8, March 2015.

[115] R. Louf and M. Barthelemy. A typology of street patterns. *Journal of The Royal Society Interface*, 11(101), 2014.

[116] openstreetmap. `http://www.openstreetmap.org/search?query=Illerrieden#map=16/48.2700/10.0514`. Accessed: 2017-02-10.

[117] Federal Motor Transport Authority (Kraftfahrt-Bundesamt KBA) Germany. Stock of passenger cars on january 1, 2016 by segment as well as by model series. `http://www.kba.de/DE/Statistik/Fahrzeuge/Bestand/`, Jan 2016.

[118] H. Schulzrinne, S. L. Casner, R. Frederick, and V. Jacobson. RTP: A Transport Protocol for Real-Time Applications. RFC 3550, July 2003.

[119] Atheros. AR9344 Highly-Integrated and Feature-Rich IEEE 802.11n 2x2 2.4/5 GHz Premium SoC for Advanced WLAN Platforms. *Data Sheet*, 2010.

[120] SKYWORKS. SE5516A: Dual-Band 802.11a/b/g/n/ac WLAN Front-End Module. *Data Sheet*, 2013.

[121] J. Bardwell. Converting Signal Strength Percentage to dBm Values. *Wild Packets Application Note*, 2002.

[122] G. Lui, T. Gallagher, B. Li, A. G. Dempster, and C. Rizos. Differences in RSSI Readings Made by Different Wi-Fi Chipsets: A Limitation of WLAN Localization. *IEEE International Conference on Localization and GNSS (ICL-GNSS)*, 2011.

[123] SKYWORKS. SE5502L: Dual-Band 802.11a/b/g/n WLAN Front-End Module. *Data Sheet*, 2012.

[124] S. Moll. Bachelorthesis: Performance Analyse von IEEE 802.11 Wireless Mesh Netzwerken – Performance Analysis of IEEE 802.11 Wireless Mesh Networks, Hochschule Ulm – University of Applied Sciences, 2015.

[125] J. Dugan. Iperf documentation. `https://iperf.fr/iperf-doc.php`. Accessed: 2015-12-06.

[126] L. M. Garcia. Manual Page TCPDUMP. `http://www.tcpdump.org/tcpdump_man.html`. Accessed: 2015-11-30.

[127] J. N. Daigle. *Queueing Theory with Applications to Packet Telecommunication*. Springer-Verlag, 2005.

[128] L. Devroye. *Non-Uniform Random Variate Generation*. Springer-Verlag, 1986.